How To Make Poultry Pay: A Practical Manual

by F.G. Paynter

with an introduction by Jackson Chambers

This work contains material that was originally published in 1907.

This publication is within the Public Domain.

*This edition is reprinted for educational purposes
and in accordance with all applicable Federal Laws.*

Introduction Copyright 2017 by Jackson Chambers

Self Reliance Books

Get more historic titles on animal and stock breeding, gardening and old fashioned skills by visiting us at:

http://selfreliancebooks.blogspot.com/

Introduction

I am pleased to present yet another title on Poultry.

The work is in the Public Domain and is re-printed here in accordance with Federal Laws.

As with all reprinted books of this age that are intended to perfectly reproduce the original edition, considerable pains and effort had to be undertaken to correct fading and sometimes outright damage to existing proofs of this title. At times, this task is quite monumental, requiring an almost total "rebuilding" of some pages from digital proofs of multiple copies. Despite this, imperfections still sometimes exist in the final proof and may detract from the visual appearance of the text.

I hope you enjoy reading this book as much as I enjoyed making it available to readers again.

Jackson Chambers

CONTENTS.

CHAPTER I.

MY ORIGINAL INVESTIGATIONS IN 1902—WHY THE FARNE ISLAND EXPERIMENT WAS UNDERTAKEN IN 1906. . 1

CHAPTER II.

RESULTS OBTAINED BY THE FARNE ISLAND EXPERIMENT MADE IN 1906 7

CHAPTER III.

THE FARNE ISLAND EXPERIMENT, 1906—THE READING EXPERIMENT, 1905—A COMPARISON OF THE TWO METHODS OF REARING CHICKENS 21

CHAPTER IV.

AN INTERESTING DEDUCTION ARRIVED AT FROM AN ANALYSIS AND COMPARISON OF THE READING AND FARNE ISLAND FIGURES 41

CHAPTER V.

APPLIANCES, FOODS, ETC., USED IN THE FARNE ISLAND EXPERIMENT 49

CONTENTS.

CHAPTER VI.

PARTICULARS OF THE FEEDING AND GENERAL MANAGEMENT OF CHICKENS ON THE FARNE ISLAND SYSTEM 61

CHAPTER VII.

THE AUTHOR'S SUGGESTIONS FOR AN EXPERIMENTAL POULTRY FARM—FACTS AND FIGURES FOR WOULD-BE POULTRY FARMERS 75

CHAPTER VIII.

A COMPARISON BETWEEN CATTLE, SHEEP, AND POULTRY FROM A FARMER'S POINT OF VIEW — THE WEIGHT OF MARKETABLE CHICKENS—HOT-AIR AND TANK INCUBATORS—CHEMICAL CONSTITUENTS OF FOODS—THE MAGNITUDE OF THE POULTRY AND EGG INDUSTRY IN THE UNITED STATES OF AMERICA . 86

HOW TO MAKE POULTRY PAY.

CHAPTER I.

MY ORIGINAL INVESTIGATIONS IN 1902. WHY THE FARNE ISLAND EXPERIMENT WAS UNDERTAKEN IN 1906.

THE poultry rearing industry ought to be one of the most profitable branches of agriculture in the British Isles, but, owing chiefly to the lack of definite and accurate information on the subject, poultry rearing does not pay.

In order to ascertain whether poultry farming could or could not be made to pay, I devoted several months in 1902 to a close practical examination of the methods in vogue in most of the South of England counties.

After a thoroughly practical investigation I came to the conclusion that poultry farming could not be made to pay if conducted by the methods taught by the Agricultural Colleges and Experimental Farms. Unfortunately these were the methods used by most of the poultry raisers with whom I came into contact, and it became quite clear to me that where profits were made they were due to special conditions quite outside the ordinary system of poultry farming. Money was certainly made by many who specialised with fancy breeds, and by those who carried on their operations in districts where commercial branches were developed, and which were situated under peculiarly favourable conditions with regard to climate, soil, general surroundings, and market facilities, such as the chicken fattening district of Heathfield, Sussex, and the duck district at and around Aylesbury. It was plain also that ordinary farmers who kept a limited stock of poultry were enabled to realise the value of their small wheat and other bye-products by turning them into chicken flesh or eggs.

As to the market for the product of the poultry farm, the following facts stand out prominently:—

The high value of chicken flesh as a food, and the enormous demand there is for it in England.

The supply of English chickens during the spring and early summer, practically throughout half the year, is entirely inadequate to the demand.

A huge sum is paid yearly for foreign poultry products.

On the middle-class table poultry is considered a luxury and not, as it might be, an ordinary change of diet.

In the United States it is estimated that on an average each person eats 26s. worth of poultry products, instead of 6s. 5d. as in England.

With these facts in view I refused to accept the conclusion that poultry farming could not be made a commercial success in this country, and I felt there must be something wrong somewhere. This view was strengthened by a chance meeting in London with an American poultry farmer, who informed me that every spring for some years back he had brought over and sold in our London markets something like 60,000 birds, and returned to the United States in July, having realised a substantial profit, although the labour, food, and other charges were greater in his country than in England.

"You English people must be asleep or blind," he remarked. "Here you have at your doors the finest markets in the world, and yet you leave them open to people such as myself to supply them. Some day," he said, "you will awake, rub your eyes, and wonder how it all happened. Then my business will be over, and if I wish to keep my English custom I shall have to make my home in England, but until then I shall continue as I am doing."

Altogether I had spent nine months in 1902, and nearly £100 upon my investigations, and although the results were not encouraging I was determined that, if ever an opportunity arose, I would make experiments alone and with my own hands in order to discover, if it were possible, the reason why English poultry farming did not pay.

The opportunity came in the spring of 1906. I therefore purchased the necessary appliances and set myself to discover the reply to this question—

What weight of the ordinary home-grown foods, as used by the English farmer, has been utilized to make one pound of chicken flesh on a bird that has arrived at a marketable weight (i.e., about four pounds live weight)?

I discovered the reply when I had reared my first brood of 41 chickens, and checked it with a second brood of 31 chickens. I had altogether under experimentation 187 chickens.

When my attention was first directed towards poultry I had sought the advice of a friend who considered himself an expert on poultry matters. His advice was to go to an agricultural college, take a diploma in agriculture or aviculture, and then try to obtain a post as a County Council lecturer on poultry matters. He said he had known people of good address who had done very well in this way.

I thought his advice sound, but remarked that supposing I went to this agricultural college, spent a year, and was taught how to make poultry farming pay in such a way that I would be capable of returning to the country to teach others how to make it a profitable branch of their local agricultural industry, then I would also be in a position to make it pay for myself, and it would be unnecessary to go as a lecturer unless I felt inclined to do so.

"Not at all," he replied. "You will know no more about practical poultry farming than you do now; but you will know more about the theory, and therefore will be able to talk to practical farmers and interest them."

"Do you mean to suggest," I asked, "that I should go round the country teaching people and urging them to take up an industry that I honestly feel has no chance of proving profitable, and that I cannot myself put into practice."

"Precisely," he answered; "and you will be doing no better or no worse than dozens of others. Because a man is a teacher or is a professor, say, of agriculture, it does not follow that he could make a farm pay."

"No," I replied, "that may be so, but it certainly is wrong to encourage people to spend their time and money on an employment the avowed object of which is to make a profit—unless one feels that with care and industry there is likely to be a reasonable reward for hard work."

I did not take the course of 12 months at the agricultural college recommended by my friend, but investigation showed me what to avoid, and experiment told me the course I must follow to make poultry rearing a commercial success. These investigations and experiments occupied nearly two years.

I have no connection whatever with any agricultural college, or vendors of poultry foods or appliances. My opinions are therefore quite unbiassed, and I have recommended foods, appliances, and methods upon sound commercial principles. Fancy poultry rearing has been entirely left alone. The following chapters show how the flesh of a four-pound chicken should be grown at a cost for food of twopence-halfpenny per pound live weight, the total cost of producing that typical chicken amounting to not more than fivepence per pound, this sum covering the cost of the egg, labour, salesman's charges and carriage, and depreciation and interest on plant. By working on the simple lines laid down, equal, if not superior, results may be obtained practically anywhere and by anyone.

There is nothing really new about my system. It is merely a combination of common-sense deductions and ordinary farmyard methods. And the truths and advantages are so very clear and obvious that I can imagine I can hear those in charge of experimental stations saying, "There is nothing new in his system at all. We have known everything he tells us for years."

My answer is that their reports show that they did not appreciate the combination, or, if they did, then they had no right to keep this knowledge to themselves.

CHAPTER II.

RESULTS OBTAINED BY THE FARNE ISLAND EXPERIMENTS IN 1906.

EGGS were placed in the incubator within a day or two of the preceding lot being hatched off. As this was continued without interruption, at the end of the fourteenth week there were five broods, containing 187 chickens, under experimentation as follows:—

No. 1 Brood, containing 41 chickens, 14 weeks of age.
,, 2 ,, ,, 31 ,, 10 ,, ,,
,, 3 ,, ,, 40 ,, 7 ,, ,,
,, 4 ,, ,, 32 ,, 4 ,, ,,
,, 5 ,, ,, 43 ,, 1 ,, ,,

187

The following are actual results, obtained week by week during the experiment, from the beginning of the first to the end of the sixteenth week. As the figures in this book are not merely scientific data, but intended for the guidance of practical workers, decimals have not been used, but all fractions have been worked out to the nearest sixteenth.

	JULY.				AUGUST.				SEPTEMBER.				OCTOBER.				Totals.
	9	16	23	30	6	13	20	27	3	10	17	24	1	8	15	22	
Number of chickens under experiment	48	46	41	77	76	74	117	117	115	150	150	150	144	187	187	187	
Pounds of A feed eaten during week	4	6	7	7	8½	13	8	10	8	7	14	15½	12	8	12½	13	
Pounds of B feed eaten during week	—	—	—	—	—	½	5½	2	5	8	—	2½	3	2½	—	1	
Pounds of C feed eaten during week	—	—	—	1	7	9	17	25	32	32	50	58	50	68	77	105	
Pounds of D feed eaten during week	—	—	—	5	10½	13½	23	38	40	47	51	70	73	65	103	109	
Pounds of food eaten during the week	4	6	7	13	26	36	53½	75	85	94	115	146	147	143½	192½	219	1362½
Cost of food eaten during the week	-/4	-/6	-/7½	-/11	1/8½	2/3½	3/1½	4/5	5/1	5/7½	6/8	8/1½	8/4½	8/1	10/9	2/2½	3 19/4
Pounds of flesh actually grown during the week	2/16	1 2/16	4 1/16	7 1/16	9 3/16	14 1/16	16 6/16	23 11/16	27 2/16	28 2/16	31 2/16	30 1/16	38	33 2/16	40 7/16	47 1/16	352 1/16
Cost per lb. (in pence) of flesh actually grown during week	8	6	1 15/16	2 4/16	2 4/16	1 5/16	2 1/16	2 1/16	2 4/16	2 4/16	2 9/16	3 1/16	2 1/16	2 1/16	3 3/16	3 1/16	2 2/16 (Average)
Pounds of food actually eaten for each lb. of flesh grown during the week	—	6	4	2	2 13/16	2 7/16	4 1/16	3 1/16	3 1/16	3 1/16	3 1/16	4 1/16	3 1/16	4 1/16	4 11/16	4 1/16	3 1/16

To obtain the total weight of the birds at the end of the experiment, the weight of the chicks when hatched (averaging 1 7/8 oz. each) must be added to the weight grown.

The above experiment started on July 2nd and ended on October 22nd. Eight chickens died during the first three weeks from eating too much dry chick feed. In consequence of this a soft chick feed was worked out, and the table given on page 56, in Chapter V., was drawn up. By following this table the fault was corrected, and in the later broods there were no deaths from this cause.

Some Facts Worth Noting.

The total weight of birds at the end of experiment was 369 lbs. 7 ozs.

The total weight of soft food eaten to produce above birds was 669 lbs., or 49 per cent.

The total weight of hard food eaten to produce above birds was $693\frac{1}{2}$ lbs., or 51 per cent.

The total weight of food eaten was $1,362\frac{1}{2}$ lbs.

The average weight of food eaten to produce a pound of flesh was $3\frac{11}{16}$ lbs.

Every pound of food eaten produced $4\frac{5}{16}$ ozs. of flesh.

Total value of food eaten, £3 19s. 4d.

The flesh produced cost per pound, live weight for food, $2\frac{9}{16}$d.

The table on the next page covers Brood No. 1 of forty-one chickens taken to sixteen weeks of age, and shows at a glance, week by week, the proportion of food eaten to flesh produced. (See also Diagram, page 20.)

First Brood: A record of 16 weeks.

	July.				August.				September.				October.				Totals.
	9	16	23	30	6	13	20	27	3	10	17	24	1	8	15	22	
Number of weeks old .	1	2	3	4	5	6	7	8	9	10	11	12	13	14	15	16	—
Number of birds . . .	49	48	41	41	41	41	41	41	41	41	41	41	41	41	41	41	—
Pounds of A feed eaten	4	6	7	7	4	9	5	3	—	—	—	—	—	—	—	—	45
Pounds of B feed eaten	—	—	—	—	—	—	—	—	—	—	—	—	—	—	—	—	—
Pounds of C feed eaten	—	—	—	1	7	9	17	24	28	27	40	40	37	28	38	48	344
Pounds of D feed eaten	—	—	1	5	$10\frac{1}{2}$	12	23	29	29	34	26	36	38	29	41	48	$361\frac{1}{2}$
Total weight of food (in pounds) eaten during week	4	6	8	13	$21\frac{1}{2}$	30	45	56	57	61	66	76	75	57	79	96	$750\frac{1}{2}$
Total cost of food eaten during week . . .	-/4	-/6	-/$7\frac{1}{2}$	-/$10\frac{3}{4}$	1/$3\frac{1}{4}$	1/11	2/$7\frac{1}{2}$	3/$1\frac{1}{2}$	3/1	3/$4\frac{1}{2}$	3/$8\frac{1}{2}$	4/$2\frac{1}{2}$	4/$1\frac{1}{2}$	3/1	4/4	5/3	42/$5\frac{1}{2}$
Cost per chicken for food (in pence) week by week	$\frac{1}{16}$	$\frac{1}{8}$	$\frac{3}{16}$	$\frac{1}{4}$	$\frac{3}{8}$	$\frac{9}{16}$	$\frac{3}{4}$	$\frac{15}{16}$	$\frac{15}{16}$	1	$1\frac{1}{16}$	$1\frac{1}{4}$	$1\frac{1}{4}$	$\frac{15}{16}$	$1\frac{1}{4}$	$1\frac{9}{16}$	—
Average weight of chickens week by week	ozs. $1\frac{9}{16}$	ozs. $2\frac{1}{16}$	ozs. $3\frac{7}{16}$	ozs. $6\frac{4}{16}$	ozs. $9\frac{8}{16}$	ozs. 14	lb oz 1 2	lb oz 1 8	lb oz 1 11	lb oz 2 4	lb oz 2 11	lb oz 3 1	lb oz 3 8	lb oz 3 11	lb oz 4 2	lb oz 4 $7\frac{1}{4}$	—

HOW TO MAKE POULTRY PAY.

The first brood was started on July 2nd and ended on October 22nd. Eight chickens died during the first three weeks from eating too much dry chick feed. The food wasted upon them has been added to the weight of food eaten by the other chickens.

	lbs.		stone	lbs.	Per stone.	s.	d.
Food eaten	45	of A dry chick feed =	3	3	at 1/2 =	3	9
,, ,,	361½	of D meal ... =	25	11½	at -/8½ =	18	3½
,, ,,	344	of C corn ... =	24	8	at -/9¾ =	19	11½
	750½		53	8¼		42	0

Therefore, 41 chickens, 16 weeks old, averaging 4 lbs. 7½ ozs. weight, should cost for food 42s., or 1s. 0¼d. per head, or 2¾d. per lb. live weight.

Therefore, 1 chicken, 16 weeks old, weighing 4 lbs. 7½ ozs., should cost to produce—

			Per lb.
Food	1/0¼	each, or	2$\frac{13}{16}$d. L.W.
Eggs and oil	-/3	,, ,,	1$\frac{1}{8}$,,
Labour	-/3	,, ,,	1$\frac{1}{8}$,,
Salesman's charges and carriage	-/3	,, ,,	1$\frac{1}{8}$,,
Depreciation and interest on plant	-/0¼	,, ,,	$\frac{1}{16}$,,
	1/9½ per head		4$\frac{3}{8}$d. L.W

Some Facts Worth Noting.

Total weight of birds at the end of experiment, 182 lbs. 13 ozs.

Total weight of food eaten to produce these birds (48 per cent. soft food and 52 per cent. hard food), 750½ lbs.

12 HOW TO MAKE POULTRY PAY.

Average weight of birds at the end of experiment, $4\frac{7\frac{1}{2}}{16}$ lbs.

Average weight of food eaten per bird, $18\frac{5}{16}$ lbs.

Weight of food eaten for each lb. of flesh produced, $4\frac{2}{16}$ lbs.

Second Brood: A Record of Twelve Weeks.

	AUGUST.				SEPTEMBER.				OCTOBER.				Totals.
	6	13	20	27	3	10	17	24	1	8	15	22	
Number of weeks old	1	2	3	4	5	6	7	8	9	10	11	12	—
Number of birds	35	33	33	33	31	31	31	31	31	31	31	31	—
Pounds of A feed eaten	$4\frac{1}{2}$	4	3	3	4	4	5	3	—	—	—	—	$30\frac{1}{2}$
Pounds of B feed eaten	—	$\frac{1}{2}$	$5\frac{1}{2}$	2	—	—	—	—	—	—	—	—	8
Pounds of C feed eaten	—	—	—	1	4	5	10	18	21	22	23	28	132
Pounds of D feed eaten	—	$1\frac{1}{2}$	—	9	11	13	18	21	19	18	32	31	$173\frac{1}{2}$
Total weight (in pounds) of food eaten during week	$4\frac{1}{2}$	6	$8\frac{1}{2}$	15	19	22	33	42	40	40	55	59	344
Total cost of food eaten during week	-/$4\frac{1}{2}$	-/$5\frac{1}{2}$	-/10	-/$10\frac{1}{2}$	1/$1\frac{1}{2}$	1/$3\frac{1}{4}$	1/11	2/$4\frac{1}{4}$	2/$2\frac{1}{4}$	2/$2\frac{1}{4}$	2/$11\frac{3}{4}$	3/2	19/9
Cost per chicken during week for food (in pence)	$\frac{1}{8}$	$\frac{3}{16}$	$\frac{5}{16}$	$\frac{3}{8}$	$\frac{7}{16}$	$\frac{1}{2}$	$\frac{3}{4}$	$\frac{7}{8}$	$1\frac{3}{16}$	$\frac{7}{8}$	$1\frac{3}{11}$	$1\frac{3}{16}$	—
Average weight of chicken week by week	oz. $2\frac{2}{16}$	oz. $3\frac{5}{16}$	oz. $5\frac{5}{16}$	oz. $8\frac{8}{16}$	oz. $12\frac{4}{16}$	lb. oz. 1 0	lb. oz. 1 6	lb. oz. 1 10	lb. oz. 1 15	lb. oz. 2 5	lb. oz. 2 9	lb.oz. $3\frac{4}{16}$	—

The above experiment started on July 30th and ended on October 22nd. Two birds were drowned during the second week during a heavy rainstorm,

HOW TO MAKE POULTRY PAY. 13

and two birds were drowned during the fifth week by falling into a rain tub.

	lbs		stone	lbs.	Per stone.	s. d.
Food eaten,	30½ of A dry chick feed	=	2	2½	at 1/2 =	2 6½
,, ,,	8 of B chick meal	=	0	8	at 1/5 =	0 9½
,, ,,	132 of C chicken corn	=	9	6	at -/9¾ =	7 8
,, ,,	173½ of D chicken meal	=	12	5½	at -/8¼ =	8 9
	344		24	8		19 9

Therefore, 31 chickens, 12 weeks old, averaging 3 lbs. weight, should cost for food 19s. 9d., or 7¾d. per head, or $2\frac{9}{16}$d. per lb. live weight, and a chicken, 12 weeks old, weighing 3 lbs., should cost to produce—

		Per lb.
Food	-/7¾ each, or	$2\frac{9}{16}$ L.W.
Eggs and oil	-/3 ,, ,,	1 ,,
Labour	-/2½ ,, ,,	$\frac{13}{16}$,,
Salesman's and carriage charges .	-/2½ ,, ,,	$\frac{13}{16}$,,
Depreciation and interest on plant .	-/0¼ ,, ,,	$\frac{1}{16}$,,
	11/4 each	$5\frac{4}{16}$d. L.W.

Some Facts Worth Noting.

Total weight of birds at end of experiment, $94\frac{9}{16}$ lbs.

Weight of food eaten to produce above birds (52 per cent. soft food and 48 per cent. hard food), 344 lbs.

Average weight of birds, $3\frac{1}{16}$ lbs.

Average weight of food eaten per bird, $11\frac{1}{16}$ lbs.

Weight of food eaten for every lb. of flesh produced, $3\frac{11}{16}$ lbs.

Third Brood: A Record of Nine Weeks.

Week	Aug. 27	September 3	10	17	24	October 1	8	15	22	Totals.
Number of weeks old	1	2	3	4	5	6	7	8	9	—
Number of birds	43	43	43	43	43	40	40	40	40	—
Pounds of A feed eaten . . .	4	4	3	6	9	9	—	—	—	35
Pounds of B feed eaten . . .	—	5	8	—	—	—	—	—	—	13
Pounds of C feed eaten . . .	—	—	—	—	—	1	18	16	29	64
Pounds of D feed eaten . . .	—	—	—	7	13	16	28	25	16	105
Total weight of food eaten during week (in pounds) . . .	4	9	11	13	22	26	46	41	45	217
Total cost of food eaten during week	-/4	-/10	1/0½	-/10	1/5	1/7½	2/5½	2/2½	2/6	13/3
Cost per chicken week by week for food (in pence) . . .	$\frac{1}{8}$	$\frac{1}{4}$	$\frac{5}{16}$	$\frac{1}{4}$	$\frac{3}{8}$	$\frac{1}{2}$	$\frac{3}{4}$	$\frac{11}{16}$	$\frac{3}{4}$	—
Average weight of chicken week by week	oz. $2\frac{4}{16}$	oz. $2\frac{14}{16}$	oz. $4\frac{1}{16}$	oz. $6\frac{4}{16}$	oz. $9\frac{3}{16}$	oz. $12\frac{9}{16}$	lb. oz. 1 2	lb. oz. 1 5	lb. oz. 1 12	—

The above experiment started on August 20th and ended on October 22nd. Three birds were killed and eaten during the sixth week by a tame gull.

	lbs.			stone	lbs.		Per stone.		s.	d.
Food eaten	35	of A dry chick feed	=	2	7	at	1/2	=	2	11
,, ,,	13	of B chick meal .	=	0	13	at	1/5	=	1	3
,, ,,	64	of C chicken corn .	=	4	8	at	-/9¾	=	3	8½
,, ,,	105	of D chicken meal .	=	7	7	at	-/8½	=	5	3½
	217								13	2

HOW TO MAKE POULTRY PAY. 15

Therefore 40 chickens, 9 weeks old, averaging 1 lb. 12 ozs. weight, cost for food 13s. 2d., or $3\frac{5}{16}$d. per head, or $2\frac{1}{4}$d. per lb.

Therefore 1 chicken, 9 weeks old, weighing 1 lb. 12 ozs., should cost to produce—

		Per lb.
Food	$3\frac{5}{16}$d. each, or	$2\frac{4}{16}$d.
Eggs and oil	3 ,, ,,	$1\frac{11}{16}$
Labour	2 ,, ,,	$1\frac{2}{16}$
Salesman's charges and carriage	$2\frac{8}{16}$,, ,,	$1\frac{8}{16}$
Depreciation and interest on plant	$\frac{4}{16}$,, ,,	$\frac{2}{16}$
	$11\frac{11}{16}$d. each	$6\frac{11}{16}$d.

Some Facts Worth Noting.

Total weight of birds at end of experiment, 70 lbs.

Total weight of food eaten to produce above birds (55 per cent. soft food and 45 per cent. hard food), 217 lbs.

Average weight of birds at end of experiment, $1\frac{12}{16}$ lbs.

Average weight of food eaten per bird, $5\frac{6}{16}$ lbs.

Weight of food eaten for each lb. of flesh produced, $3\frac{2}{16}$ lbs.

How the Cost was Calculated.

Egg Charges, $2\frac{1}{2}$d. per Bird.

The average price of new-laid eggs in our

market towns has been ascertained to be as follows:—

	s. d.			s. d.	
January .	. 1 8	per doz.	July .	. 0 11	per doz.
February	. 1 6	,,	August .	. 1 2	,,
March .	. 1 0	,,	September	. 1 4	,,
April	. 0 10	,,	October .	. 1 6	,,
May	. 0 9	,,	November	. 1 8	,,
June .	. 0 10	,,	December	. 1 10	,,

Average cost of new-laid eggs = 1s. 3d. per doz., or 1¼d. each.

Two eggs were purchased for every bird that was reared and sold. The wastage in 100 eggs averaged as follows:—twenty unfertile; five misshapen, broken, or cracked; twenty failed to hatch; five birds died before coming to maturity. Therefore, to produce a chicken two eggs were required at 1¼d. each = 2½d.

Oil Charges, ½d. per Bird.

One incubator cost about 4d. per week for oil, and two foster mothers cost about 1½d. per week each for oil. As these were producing about fourteen chickens per week, the charge against each bird for this item worked out at about ½d. per head.

Labour Charges, 3d. per Bird.

I estimated that one man at £1, and one boy at 5s. per week, can efficiently look after and produce 100 chickens per week; that is, they would have under

HOW TO MAKE POULTRY PAY.

their charge 1,600 chickens in sixteen lots of 100 each, ranging from one to sixteen weeks old.

This will represent a charge of 3d. per head on each bird reared.

Charges for Salesman and Carriage, 3d. per Bird.

The salesman's commission is 5 per cent. Therefore, on birds returning an average of 2s. 8d. per head, the charge will be $1\frac{5}{8}$d. per bird. This leaves $1\frac{3}{8}$d. per head for carriage. This charge will, of course, depend upon the distance between the farm and the market town.

How the cost of the food was worked out is shown in a later chapter, where the vitally important rules for selection and mixing are laid down.

From the above it will be seen that—

No. of Chickens.	Weeks old.		lbs. ozs.		s. d.		Per lb.	
41	16	averaged	4 7	had cost	1 $0\frac{1}{4}$	each, or	$2\frac{13}{16}$d.	L.W.
72	12	in	3 $0\frac{1}{2}$	for	0 $7\frac{5}{8}$	,,	$2\frac{6}{10}$d.	,,
112	9	weight	1 14	food	0 $4\frac{3}{10}$	,,	$2\frac{3}{10}$d.	,,

I have shown what it costs to produce chickens. The following table will show us what we may expect to obtain for them in our provincial towns and in London. The provincial chickens are estimated to weigh 4 lbs. each when marketed, and to be hand fed throughout. The London chickens are estimated to weigh $5\frac{1}{2}$ lbs. each, and to be finished by cramming in the ordinary way. If the birds should weigh less

c

18 HOW TO MAKE POULTRY PAY.

than the above they may bring less, but if more, they may be expected to realise more in proportion.

The following is just a fair all-round market price for young chickens. The provincial birds are marketed alive, but the London birds are killed and prepared according to the Surrey style.

Market Prices of a Bird in the Provinces Weighing Four Pounds.

	JAN.	FEB.	MAR.	APL.	MAY	JUNE	JULY	AUG.	SEPT.	OCT.	NOV.	DEC.
Price per bird	s. d. 2 6	s. d. 2 9	s. d. 3 0	s. d. 3 3	s. d. 3 3	s. d. 3 0	s. d. 2 9	s. d. 2 3	s. d. 2 3	s. d. 2 3	s. d. 2 3	s. d. 2 6
Price per pound	0 7½	0 8¼	0 9	0 9¾	0 9¾	0 9	0 8¼	0 6¾	0 6¾	0 6¾	0 6¾	0 7½

Average price per bird, 2s. 8d. Average price per pound, 8d.

Market Prices of a Bird in London Weighing Five and a Half Pounds.

	JAN.	FEB.	MAR.	APL.	MAY	JUNE	JULY	AUG.	SEPT.	OCT.	NOV.	DEC.
Price per bird	s. d. 4 0	s. d. 4 3	s. d. 4 6	s. d. 5 6	s. d. 6 0	s. d. 5 6	s. d. 4 0	s. d. 3 6	s. d. 3 6	s. d. 3 6	s. d. 3 0	s. d. 3 6
Price per pound	0 9	0 9½	0 10	1 0	1 1	1 0	0 9	0 7¾	0 7¾	0 7¾	0 7¼	0 7¾

Average price per bird, 4s. 2¾d. Average price per pound, 9¼d.

HOW TO MAKE POULTRY PAY. 19

As cramming, preparing, and bringing the birds up to the London market standards of weight, quality, and appearance, will add an additional cost of 1s. per bird, it would appear from the above that it is not (as is generally supposed) so very much more profitable to cater for the London market than for the provincial.

A 4½-lb. chicken which will have cost, if produced for a provincial market (as explained), 1s. 9½d., will realise an average of 8d. per lb., or an average price each of 3s., leaving an average profit of 1s. 2½d. per bird.

Had this bird been crammed and prepared for the London market, it would then have weighed 6 lbs., and would have cost for extra food, killing, dressing, etc., probably about an additional 1s.

The cost of production would then have been 2s. 9½d., and it might have been expected to realise 9¼d. per lb., or an average price of 4s. 7½d., leaving an average profit of 1s. 10d. per bird.

The following diagram refers to Brood No. 1 of forty-one chickens taken to sixteen weeks of age, and shows at a glance, week by week, the proportion of food eaten to the flesh produced:—

Explanation of diagram on the following page.

NOTE.—The figures on top line show the amount of food eaten each week. Grand total, 750 pounds. The vertical lines mark the weeks, 1 to 16. The shaded portion shows how much flesh was added, week by week. The line A B represents one pound of food, which is divided into ounces on the scale E F. For example, during the fourth week every pound of food was turned into 8 ounces of live weight. This is shown by point H, which makes

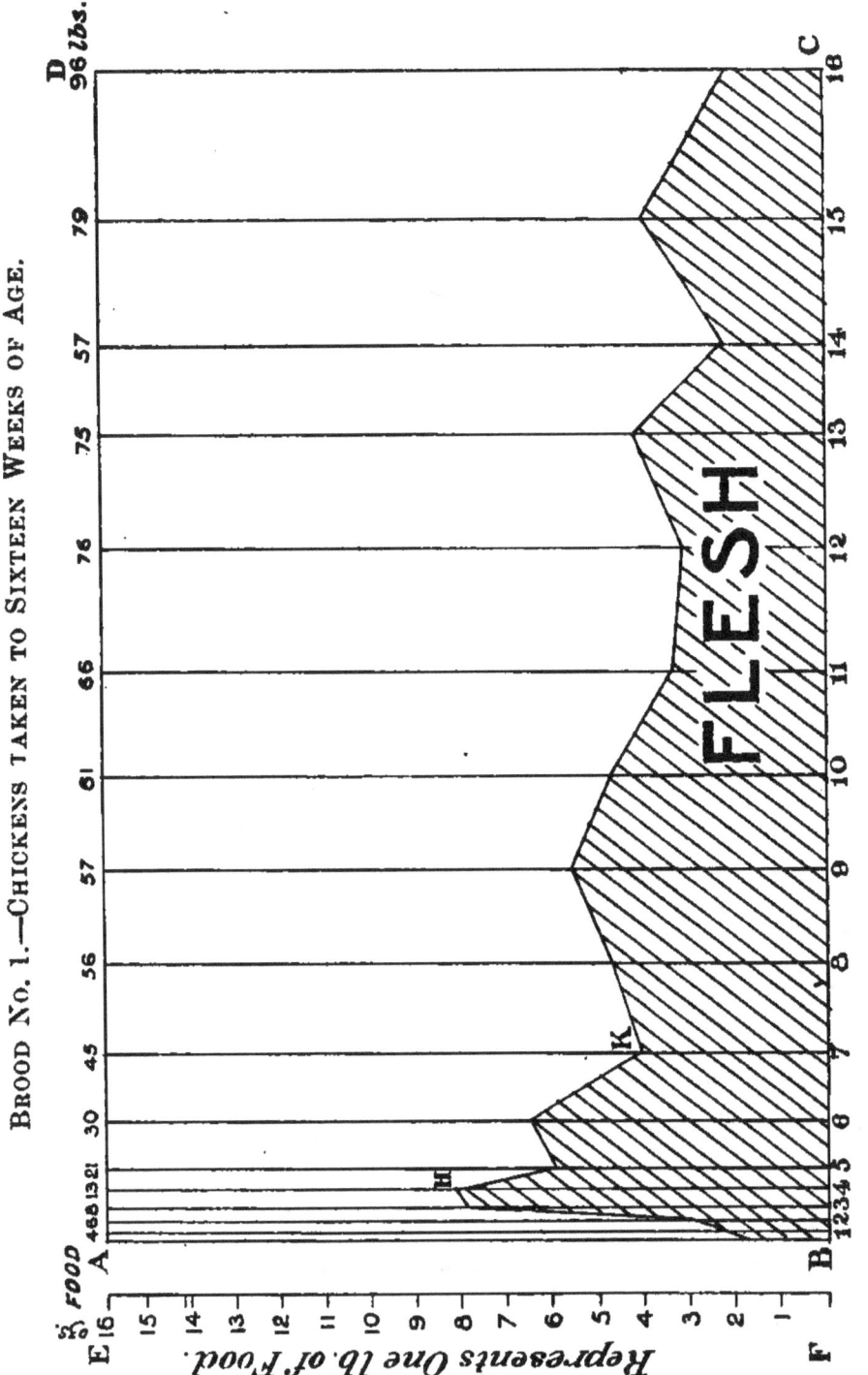

BROOD No. 1.—CHICKENS TAKEN TO SIXTEEN WEEKS OF AGE.

it clear that it required two pounds of food to make one pound of live weight. The birds during this week consumed 13 pounds of food, which is shown by the figures on the top line. Again, during the seventh week every pound of food was turned into 4 ounces of live weight, see point K. During this week, therefore, four pounds of food made one pound of live weight.

CHAPTER III.

THE FARNE ISLAND EXPERIMENT, 1906. THE READING EXPERIMENT, 1905. A COMPARISON OF THE TWO METHODS OF REARING CHICKENS.

THE following pages are devoted to a close comparison between the Farne Island experiments made in 1906, and an experiment conducted by the Reading College Poultry Farm, Theale, in 1905, on the growth and cost of rearing chickens on the lines laid down and advocated by that institution. In my own experiment on Farne Island one object was held in view—namely, the discovery and careful record of methods of profitable poultry raising under quite ordinary circumstances.

General Conditions under which the Two Experiments were conducted.

	READING.	FARNE ISLAND.
Locality . .	Theale, Berkshire. Thames Valley.	Farne Islands, off the Northumberland Coast.
Time of Year	From March to July, the natural breeding and growing season.	From July to Oct., an unnatural breeding and growing season.

	READING.	FARNE ISLAND.
Breed . . .	Cross Breeds, Buff Orpington and Houdan, Buff Orpington and Indian Game.	Barndoor Fowls, Crossbreeds, Plymouth Rock, Leghorn, etc. In the first brood only there was some game blood.
Eggs . . .	New-laid. Self produced.	Bought haphazard from higglers, shops, and small farmers.
Housing . .	Large portable houses, worth probably about £4 each, with 7·8 c.f. air space per bird.	Small self-made houses, worth from 20s. to 25s. each, with 2·32 c.f. air space per bird.
Attendance .	Under Mr. E. T. Brown and Mr. Will Brown, and under the personal superintendence of Mr. Edward Brown.	F. G. Paynter only.
General Conditions of Rearing and Feeding	Reared on the most advanced principles, and fed in the best manner known to us. (Signed) EDWARD BROWN, E. T. BROWN, WILL BROWN.	Reared and fed in the best manner known to me. (Signed) F. G. PAYNTER.

Foods used.

	READING.	FARNE ISLAND.
Dry Chick Feed	*A Mixture.* BY WEIGHT. Wheat, Cracked 3 parts. Dari . . . . 2 ,, Canary Seed . . 2 ,, Oatmeal, pinhead . . . 2 ,, Millet . . . . 2 ,, Broken Maize . 1 ,, Hemp Seed . . 1 ,,	*A Mixture.* BY WEIGHT. Wheat, Cracked 3 parts. Dari . . . . 2 ,, Canary Seed . . 2 ,, Oatmeal . . . 2 ,, Millet . . . . 1 ,, Broken Maize . 1 Buckwheat (small) . . . 1 ,,

HOW TO MAKE POULTRY PAY.

	READING.	FARNE ISLAND.
Dry Chick Feed (continued)	*A Mixture.* BY WEIGHT. Rice . . . . 1 part Meat . . . 1 ,, Grit . . . . 1 ,, Cost 10s. 8d. per cwt., or 1$\frac{2}{10}$ per lb. Protein 12; fat, 6; carbon, 62; bone-making, 2; fibre, 2; water, 16—Total, 100.	*A Mixture.* BY WEIGHT. Rice . . . . 1 part Meat . . . 1 ,, Grit . . . . 1 ,, Cost, 9s. 4d. per cwt., or 1d. per lb. Protein, 12; fat, 4; carbon, 65; bone-making, 2; fibre, 2; water, 15—Total, 100.
Chicken Corn	*B Mixture.* BY WEIGHT. Wheat, Cracked 3 parts. Broken Maize . 2 ,, Dari . . . . 2 ,, Buckwheat . . 2 ,, Rice . . . . 1 ,, Hemp Seed . . 1 ,, Meat . . . . 1 ,, Linseed . . . 1 ,, Grit and Oyster Shell . . . 1 ,, Cost, 7s. 6d. per cwt., or $\frac{13}{16}$d. per lb. Protein, 12; fat, 9; carbon, 59; bone-making, 2; fibre, 3; water, 15—Total, 100.	*C Mixture.* Wheat . . 25 per cent. Barley . . 25 ,, Maize . . . 20 ,, Dari . . . 10 ,, Buckwheat . 10 ,, Tares . . 10 ,, Cost, 6s. 6d. per cwt., or $\frac{11}{16}$d. per lb. Protein, 12; fat, 4; carbon, 65; bone-making, 2; fibre, 6; water, 11—Total, 100.
Chicken Meal	*D Mixture.* Spratts' Patent Chicken Meal. Cost, 18s. 4d. per cwt., or 2d. per lb. Protein, 17·44; fat, 1·98; carbon, 67·23; fibre, 0·46; water, 9·72; mineral matter, 3·17—Total, 100.	*B Mixture.* BY BULK. Biscuit Meal . 6 parts. D Chicken Meal 2 ,, Rice . . . . 2 ,, Oatmeal . . . 2 ,, Meat . . . . 1 ,, Cost, 11s. 4d. per cwt., or 1$\frac{3}{10}$d. per lb. Protein, 13$\frac{1}{2}$; fat, 2; carbon, 66$\frac{1}{2}$; bone-making, 2; water, 15; fibre, 1—Total, 100.

	READING.	FARNE ISLAND.
Chicken Meal	*C Mixture.* BY WEIGHT. Barley Meal . . 4 parts. Toppings . . . 4 „ Meat . . . . 1 „ Cost, 7s. 9d. per cwt., or 1¾d. per lb. Protein, 15 ; fat, 3 ; carbon, 52 ; bone-making, 4 ; water, 18 ; fibre, 8 —Total, 100.	*D Mixture.* BY WEIGHT. Middlings . . 3 parts. Coarse Bran . 1 „ Cost, 5s. 8d. per cwt., or ⅝d. per lb. Protein, 17 ; fat, 5 ; carbon, 51 ; bone-making, 8 ; water, 14 ; fibre, 5 —Total, 100.
Corn . .	*E Mixture.* Wheat. Cost, 7s. per cwt., or ¾d. per lb. Protein, 12 ; fat, 3 ; carbon, 70 ; bone-making, 2 ; water, 12 ; fibre, 1 —Total, 100.	

Foods Used, with Methods and Time of Feeding.

	READING.	FARNE ISLAND.
From 1 to 10 days	The chicks are apparently given an unlimited amount of dry chick food.	The chicks are given a limited amount of A dry chick feed. Fed to them systematically by Table.
From 10 days to 14 days	Same as above.	The chicks now have two additional feeds of C soft food mixture at 9 and 3.
From 14 days to 28 days	Same as above.	The chicks now have five regular meals per day, the A mixture at daybreak, 12, and sunset and the C soft food at 9 and 3.

	READING.	FARNE ISLAND.
From 4 weeks to 8 weeks	They are given four meals per day; D mixture at 7 a.m. and 2.30 p.m., and the B grain mixture at 10 a.m. and 5 p.m.	They are given four meals per day; D soft-food mixture at 9 a.m. and 3 p.m., and C grain mixture at 12 p.m. and sunset.
From 8 weeks to marketing	C soft-food mixture at 7 a m. E grain mixture at 12 p.m. and 5 p.m.	Same as above.
	Just fed in the ordinary way throughout.	Fed systematically by Table as described in previous chapter throughout.
	During the experiment 216 lbs. of food were eaten.	During the experiment 518 lbs. of food were eaten.
	Cost, 14s. 9d., or 7s. 7d. per cwt., or 1⅜d. per lb.	Cost, 29s. 9d., or 6s. 5d. per cwt., or 1⅛d. per lb.

A Comparison of Results.

	READING.	FARNE ISLAND.
Incubation	75% of fertile eggs.	79% of fertile eggs.
Egg Expenses	1$\frac{7}{10}$d. per chicken.	2½d. each. The unfertile eggs are included, and are charged for. (See page 15.)
	The initial cost of eggs was charged at 1s. per dozen.	The initial cost of eggs was charged at 1s. 3d. per dozen.
Oil	⅜d. per chicken.	½d. per chicken.

A Brood of Thirty Chickens.
Taken to Thirteen Weeks of Age at Reading in 1905.

Week ending	APRIL				MAY					JUNE				JULY	Totals
	12	19	26	3	10	17	24	31	7	14	21	28	5		
Number of weeks old	1	2	3	4	5	6	7	8	9	10	11	12	13	
Number of birds	30	30	30	30	30	30	30	30	30	30	30	30	30	
Pounds of A feed eaten	2¼	6¼	8½	7¼	—	—	—	—	—	—	—	—	—	24¼
Pounds of B feed eaten	—	—	—	—	9¼	16¾	17¾	18½	—	—	—	—	—	62¼
Pounds of C feed eaten	—	—	—	—	—	—	—	—	2¼	3	3½	3¼	4	16
Pounds of D feed eaten	—	—	—	—	¾	1¼	1½	2	—	—	—	—	—	5½
Pounds of E feed eaten	—	—	—	—	—	—	—	—	18¾	18	19½	19	19¾	95
Total weight of food eaten during the week (in pounds)	2¼	6¼	8½	7¼	10	18	19¼	20½	21	21	23	22¼	23¾	203
Total cost of the food eaten during the week	-/2¼	-/6¾	-/9½	-/7¾	-/9	1/4	1/5	1/7	1/4	1/4	1/5½	1/5½	1/6¼	14/4½
Cost for food per chicken week by week	-/1⅕	-/1⅜	-/1⁵⁄₁₆	-/1⁴⁄₁₆	-/1⅜	-/1⅜	-/1⅜	-/1⅜	-/1⅜	-/1⅜	-/1⅜	-/1⅜	-/1⅜	-/1⅜
Average weight of chickens week by week	ozs.	ozs.	ozs.	ozs.	ozs. lb oz	ozs. lb oz	ozs. lb oz	ozs. lb oz	ozs. lb oz	ozs. lb oz	ozs. lb oz	ozs. lb oz	ozs. lb oz	
	1⅞	2⁶⁄₁₆	3⁵⁄₁₆	5⁹⁄₁₆	7⁵⁄₁₆	9	10¾ 1 0 1	10¼ 1 0 1	3 1 6 1	7 1 10 1	13			
Average weight of food eaten per chicken week by week	1⅛	3⅜	4⁶⁄₁₆	3½	5⁵⁄₁₆	9⅜	9¾	10¾	11⅜	11⅞	12⁴⁄₁₆	11¹⁴⁄₁₆	12¹⁰⁄₁₆	

HOW TO MAKE POULTRY PAY.

The above experiment started on April 12th and ended on July 5th, 1905. No chickens died during the experiment.

	lbs.			stone lbs.	Per stone.	s.	d.
Food eaten	24	of A dry chick feed	=	1 9¾	at 1/4	2	3
,, ,,	62¼	of B corn	=	4 6	at -/11¼	4	2
,, ,,	16	of C meal	=	1 1½	at -/11⅒	1	1
,, ,,	5½	of D meal	=	0 5½	at 2/3½	0	11
,, ,,	95	of E corn	=	6 11	at -/10½	5	11
	201¼					14	4

Therefore, 30 chickens, 13 weeks old, averaging 1 lb. 13 ozs. weight, cost for food, 14s. 4d., or 5$\frac{12}{16}$d. per head, or 3$\frac{3}{16}$ per lb. live weight.

Therefore, 1 chicken, 13 weeks old, weighing 1 lb. 13 ozs., will cost, if produced by the Reading system :—

		Per lb.
Per head for Food	-/5$\frac{12}{16}$ each, or	-/3$\frac{3}{16}$ L.W.
,, Eggs and oil	-/3 ,,	,, -/1$\frac{1}{16}$,,
,, Labour	-/3 ,,	,, -/1$\frac{1}{16}$,,
,, Salesman's charges and carriage	-/3 ,,	,, -/1$\frac{1}{16}$,,
,, Depreciation and interest on plant	-/0¼ ,,	,, -/0$\frac{2}{16}$,,
	1/3 each	-/8$\frac{5}{16}$ L.W.

Some Facts Worth Noting.

Total weight of birds at end of experiment, 56 lbs.

Total weight of food eaten to produce above birds, 201 lbs.

Ninety per cent. of the food eaten was hard, and ten per cent. soft food.

Brood of Forty-one Chickens.

Taken to Thirteen Weeks of Age at the Farne Islands in 1906.

														Totals.
Number of weeks old . . .	1	2	3	4	5	6	7	8	9	10	11	12	13	—
Number of birds	49	48	41	41	41	41	41	41	41	41	41	41	41	—
Pounds of A feed eaten . . .	4	6	7	7	4	9	5	3	—	—	—	—	—	45
Pounds of B feed eaten . . .	—	—	—	—	—	—	—	—	—	—	—	—	—	—
Pounds of C feed eaten . . .	—	—	—	1	7	9	17	24	28	27	40	40	37	230
Pounds of D feed eaten . . .	—	—	1	5	$10\frac{1}{2}$	12	23	29	29	34	26	36	38	$243\frac{1}{2}$
Total weight of food eaten during the week (in pounds)	4	6	8	13	$21\frac{1}{2}$	30	45	56	57	61	66	76	75	518
Total cost of food eaten during the week	-/4	-/6	-/$7\frac{1}{2}$	-/$10\frac{3}{4}$	1/$3\frac{1}{4}$	1/$10\frac{3}{4}$	2/7	3/$1\frac{1}{2}$	3/1	3/$3\frac{3}{4}$	3/$7\frac{3}{4}$	4/2	4/$1\frac{1}{4}$	29/$6\frac{1}{2}$
Cost for food per chicken week by week	-/$\frac{1}{16}$	-/$\frac{1}{8}$	-/$\frac{3}{16}$	-/$\frac{1}{4}$	-/$\frac{3}{8}$	-/$\frac{9}{16}$	-/$0\frac{3}{4}$	-/$\frac{15}{16}$	-/$\frac{15}{16}$	-/1	/$1\frac{1}{16}$	-/$1\frac{1}{4}$	-/$1\frac{1}{4}$	—
Average weight of chickens week by week	ozs. $1\frac{10}{16}$	ozs. $2\frac{1}{16}$	ozs. $3\frac{7}{16}$	ozs. $6\frac{4}{16}$	ozs. $9\frac{6}{16}$	ozs. 14	lb oz 1 2	lb oz 1 8	lb oz 1 11	lb oz 2 4	lb oz 2 11	lb. oz. 3 1	lb. oz. 3 8	—
Average weight of food eaten per chicken week by week .	ozs. $1\frac{5}{16}$	ozs. $2\frac{1}{16}$	ozs. $3\frac{2}{16}$	ozs. $5\frac{1}{16}$	ozs. $8\frac{6}{16}$	ozs. $11\frac{11}{16}$	ozs. $17\frac{9}{16}$	ozs. $21\frac{9}{16}$	ozs. $22\frac{4}{16}$	ozs. $23\frac{12}{16}$	ozs. $25\frac{12}{16}$	ozs. $29\frac{10}{16}$	ozs. $29\frac{4}{16}$	—

HOW TO MAKE POULTRY PAY. 29

At the end of the experiment, shown on the opposite page, the average weight of birds was 29 ozs.

Average weight of food eaten per bird, $6\frac{1}{16}$ ozs.

Every pound of food eaten produced $4\frac{7}{16}$ ozs. of flesh.

The grit and litter used are not taken under consideration, as the value of the manure made will more than cover these charges.

The above experiment started on July 2nd and ended on October 1st, 1906. Eight chickens died during the first three weeks from eating too much dry chick feed.

	lbs.		stone	lbs.	Per stone.	s.	d.
Food eaten	45	of A dry chick feed =	3	3	at 1/2 =	3	9
,, ,,	230	of C corn . . . =	16	6	at -/9¾ =	13	4
,, ,,	243½	of D meal . . . =	17	9½	at -/8½ =	12	6
	518½					29	7

Therefore, 41 chickens, 13 weeks old, averaging 3 lbs. 8 ozs. weight, cost for food, 29s. 7d., or 8¾d. per head, or $2\frac{7}{16}$d. per lb. live weight.

Therefore, 1 chicken, 13 weeks old, weighing 3 lbs. 8 ozs., will cost, if produced by the Farne Island system :—

		Per lb.
Per head for Food	-/8¾ each, or	-/2$\frac{7}{16}$ L.W.
,, Eggs and oil	-/3 ,, ,,	-/1$\frac{1}{8}$,,
,, Labour	-/3 ,, ,,	-/1$\frac{1}{8}$,,
,, Salesman's charges and carriage .	-/3 ,, ,,	-/1$\frac{1}{16}$,,
,, Depreciation and interest on plant	-/0¼ ,, ,,	-/$\frac{1}{16}$,,
	1/6 each.	-/5$\frac{1}{8}$ L.W.

HOW TO MAKE POULTRY PAY.

Some Facts Worth Noting.

Total weight of birds at end of experiment, 145 lbs.

Total weight of food eaten to produce above birds, 516 lbs.

Fifty-five per cent. of the food eaten was hard, and forty-five per cent. soft food.

Average weight of birds at end of experiment, 56 ozs.

Average weight of food eaten per bird, $12\frac{9}{16}$ ozs.

Every pound of food eaten produced $4\frac{8}{16}$ ozs. of flesh.

The grit and litter used are not taken under consideration, as the value of the manure will more than cover these charges.

Chicken manure may be worth anything from £1 to £4 per ton. The price will depend on the quality and the local demand. If it is to be used for tomato growing it may fetch £4 per ton, but if for grass-land then £1 would be a fair price.

A Comparison of Results at Equal Weights and Equal Ages.

	Number of birds.	Age in weeks.	Average weight (in ounces).	Cost for food per bird (in pence).	Cost per lb. L.W. for food (in pence).	Number of birds.	Age in weeks.	Average weight (in ounces).	Cost for food per bird (in pence).	Cost per lb. L.W. for food (in pence).
Reading	30	13	29	$5\frac{2}{10}$	$3\frac{3}{10}$	30	13	29	$5\frac{2}{10}$	$3\frac{3}{10}$
Farne Island	41	13	56	$8\frac{3}{4}$	$2\frac{7}{10}$	41	9	31	$4\frac{3}{10}$	$2\frac{2}{10}$

A Comparison from a Financial Point of View.

READING.

Credit.	s.	d.	Debit.	s.	d.
One Chicken, 13 weeks old, weighing 29 ozs., at 8d. per lb.	1	$2\frac{8}{16}$	Food	0	$5\frac{12}{16}$
			Eggs and Oil	0	3
			Labour	0	3
			Salesman's Charges and Carriage	0	3
			Depreciation and Interest	0	$0\frac{4}{16}$
Loss	0	$1\frac{5}{16}$	Rent	0	$0\frac{3}{16}$
	1	$3\frac{13}{16}$		1	$3\frac{13}{16}$

FARNE ISLAND

Credit.	s.	d.	Debit.	s.	d.
One Chicken, 13 weeks old, weighing 56 ozs., at 8d. per lb.	2	4	Food	0	$8\frac{12}{16}$
			Eggs and Oil	0	3
			Labour	0	3
			Salesman's Charges and Carriage	0	3
			Depreciation and Interest	0	$0\frac{4}{16}$
			Rent	0	$0\frac{3}{16}$
			Profit	0	9
	2	4		2	4

The point might be raised that it is not practical to market chickens so small or so young as from 12 to 13 weeks old.

Reading, however, made a somewhat similar experiment in 1904. Their chickens then averaged in weight, at 12 weeks, $2\frac{5}{16}$ lbs. They were sold for table purposes through the ordinary trade channels, and realised what appears to me to be rather a remarkably good price, namely, 1s. 4d. per lb. live weight.

Question of Rent.

As above comparison of results is intended to refer to poultry rearing on a large practical scale, year in and year out, the question of rent must be taken under consideration. The droppings of 205 birds reared on one acre amount to a fair manurial dressing to the land, so that one acre of grass-land may safely be used indefinitely for rearing this number of chickens per year without fear of injury.

Therefore, every 205 birds will require one acre of good grass-land, which at £2 per acre represents a charge of $1\frac{3}{16}$d. per head.

But as my system is entirely a non-wire system, there being no wire runs, and the houses moved to a fresh spot every time a brood is marketed—that is, every 10 weeks—it will be possible to graze the land during the late spring, summer, and autumn with Irish or Scotch cattle, and thus considerably assist with the rent.

I have proved in the previous chapter that one may reasonably expect an average market return of 8d. per lb. live weight all the year round.

HOW TO MAKE POULTRY PAY. 33

Therefore from the above it appears that whereas the Farne Island system is distinctly profitable, the Reading system (which is that taught by the agricultural colleges in this country and in general use here) is distinctly unprofitable.

No matter how picturesque a technical teaching may be, it is commercially valueless unless when put to a practical test a profit can be shown. If it cannot do this, then it must bring ruin to those entering it with the expectation of making a livelihood.

Unless the average market value of the flesh or eggs produced week by week (whichever the object may be) is sufficient to pay for the average cost of food eaten, rent and all charges, and leave a margin for profit, the industry has no foundation, and it is neither commercial or valuable. As this point is of the utmost importance, I have, in order to prove the commercial aspect of the two systems, drawn up the following statements comparing the progress of the two broods week by week side by side, and then, in order that the lesson learnt may be still more clearly seen, have expressed the results by a diagram.

One great advantage obtained by the Farne Island system over that taught at Reading is that it makes the best use of the period of a chicken's life when flesh is grown at the most economic rate—that is between the fourth and the thirteenth week. (*See* Diagram, on page 20) Reading loses this opportunity and, therefore, loses the profit.

D

READING AVERAGE.

Number of weeks old	1	2	3	4	5	6	7	8	9	10	11	12	13
Increased weight of flesh produced by each chicken	$\tfrac{5}{16}$	$\tfrac{5}{16}$	$1\tfrac{4}{16}$	$2\tfrac{4}{16}$	$1\tfrac{12}{16}$	$1\tfrac{11}{16}$	$2\tfrac{1}{16}$	$6\tfrac{2}{16}$	$2\tfrac{4}{16}$	$3\tfrac{5}{16}$	$1\tfrac{4}{16}$	3	$3\tfrac{3}{16}$
Value of increased flesh at 8d. per pound	$\tfrac{2}{16}$	$\tfrac{3}{16}$	$\tfrac{7}{16}$	$1\tfrac{2}{16}$	$1\tfrac{4}{16}$	$1\tfrac{4}{16}$	$1\tfrac{6}{16}$	$3\tfrac{1}{16}$	$1\tfrac{2}{16}$	$1\tfrac{10}{16}$	$\tfrac{10}{16}$	$1\tfrac{8}{16}$	$1\tfrac{9}{16}$
Value of food eaten by each chicken during the week	$\tfrac{1}{16}$	$\tfrac{3}{16}$	$\tfrac{5}{16}$	$\tfrac{4}{16}$	$\tfrac{5}{16}$	$\tfrac{9}{16}$	$\tfrac{9}{16}$	$1\tfrac{1}{16}$	$\tfrac{8}{16}$	$\tfrac{9}{16}$	$\tfrac{9}{16}$	$\tfrac{9}{16}$	$1\tfrac{0}{16}$
Margin of profit left after paying for the food eaten	$\tfrac{1}{16}$	$\tfrac{1}{16}$	$\tfrac{2}{16}$	$1\tfrac{4}{16}$	$\tfrac{9}{16}$	$\tfrac{5}{16}$	$\tfrac{13}{16}$	$2\tfrac{6}{16}$	$\tfrac{10}{16}$	$1\tfrac{1}{16}$	$\tfrac{1}{16}$	$\tfrac{13}{16}$	1
Total profit to date on each bird	$\tfrac{1}{16}$	0	$\tfrac{2}{16}$	1	$1\tfrac{9}{16}$	$1\tfrac{14}{16}$	$2\tfrac{11}{16}$	$5\tfrac{1}{16}$	$5\tfrac{11}{16}$	$6\tfrac{12}{16}$	$6\tfrac{13}{16}$	$7\tfrac{13}{16}$	$8\tfrac{13}{16}$

FARNE ISLAND AVERAGE.

Number of weeks old	1	2	3	4	5	6	7	8	9	10	11	12	13
Increased weight of flesh produced by each chicken	$\tfrac{3}{16}$	$\tfrac{7}{16}$	$1\tfrac{6}{16}$	$3\tfrac{3}{16}$	$3\tfrac{2}{16}$	$4\tfrac{10}{16}$	$4\tfrac{4}{16}$	$5\tfrac{12}{16}$	$7\tfrac{1}{16}$	$6\tfrac{3}{16}$	$5\tfrac{4}{16}$	$5\tfrac{3}{16}$	$7\tfrac{7}{16}$
Value of increased flesh at 8d. per pound	$\tfrac{2}{16}$	$\tfrac{3}{16}$	$1\tfrac{1}{16}$	$1\tfrac{10}{16}$	$1\tfrac{9}{16}$	$2\tfrac{5}{16}$	$2\tfrac{2}{16}$	$2\tfrac{14}{16}$	$3\tfrac{11}{16}$	$3\tfrac{7}{16}$	$2\tfrac{10}{16}$	$2\tfrac{10}{16}$	$3\tfrac{11}{16}$
Value of food eaten by each chicken during the week	$\tfrac{1}{16}$	$\tfrac{2}{16}$	$\tfrac{3}{16}$	$\tfrac{4}{16}$	$\tfrac{6}{16}$	$\tfrac{9}{16}$	$\tfrac{12}{16}$	$\tfrac{15}{16}$	$\tfrac{15}{16}$	1	$1\tfrac{1}{16}$	$1\tfrac{1}{16}$	$1\tfrac{4}{16}$
Margin of profit left after paying for the food eaten	$\tfrac{1}{16}$	$\tfrac{1}{16}$	$\tfrac{5}{16}$	$1\tfrac{6}{16}$	$1\tfrac{3}{16}$	$1\tfrac{12}{16}$	$1\tfrac{6}{16}$	$1\tfrac{15}{16}$	$2\tfrac{15}{16}$	$2\tfrac{7}{16}$	$1\tfrac{9}{16}$	$1\tfrac{6}{16}$	$2\tfrac{7}{16}$
Total profit to date on each bird	$\tfrac{1}{16}$	$\tfrac{2}{16}$	$1\tfrac{1}{16}$	$2\tfrac{1}{16}$	$3\tfrac{4}{16}$	5	$6\tfrac{7}{16}$	$8\tfrac{6}{16}$	$11\tfrac{5}{16}$	$13\tfrac{13}{16}$	$15\tfrac{7}{16}$	$16\tfrac{13}{16}$	$19\tfrac{4}{16}$

Diagram showing the balance of profit standing to the credit of chickens reared during the Farne Island experiment in 1906, and chickens reared during the Reading experiment in 1905, after the value of the food they had eaten had been deducted from the value of the flesh produced at 8d. per pound live weight. From this balance all charges must be paid before profits may be estimated. Shaded columns represent the balance standing to the credit of the Farne Island birds at the end of each week, and the plain columns show the standing of the Reading birds.

It was found that after the thirteenth week the proportion of flesh produced per pound of food eaten was much lower. For instance, at the Farne Islands up to the end of the thirteenth week each pound of food eaten produced on an average $4\frac{8}{16}$ ozs. of flesh; but between the thirteenth and sixteenth week each pound eaten only produced $3\frac{2}{6}$ ozs., or 18 per cent. less.

It was, therefore, found that the older the birds grew the more expensive was their flesh to produce, as follows:—

From the 1st to the 6th week the flesh during the Farne Island experiment cost to produce for food, $1\frac{4}{16}$d. per lb.

At the 9th week the flesh was costing to produce for food $3\frac{4}{16}$d. per lb.

At the 12th week the flesh was costing to produce for food $2\frac{5}{8}$d. per lb., or 10 per cent. less.

At the 15th week the flesh cost to produce for food $3\frac{6}{16}$d. per lb.

At the 16th week the flesh was costing to produce for food $4\frac{6}{16}$d. per lb., or about 30 per cent. more.

Why the Farne Island system is so much more profitable than that taught at Reading is attributed to the following reasons:—

The large proportion of soft food used, 45 per cent. of the food eaten by the Farne Island birds being soft food, against only 10 per cent. at Reading, and as $77\frac{1}{2}$ per cent. of poultry flesh is water, and 75 per cent. of that of most animals, therefore unless growing animals can obtain what nature demands and seeks to absorb, their growth must be retarded. Probably in the case of crammed birds

HOW TO MAKE POULTRY PAY.

more than 90 per cent. of the extra weight gained is purely water.

The analysis of poultry flesh is: Water, 77½; nitrogenous matter, 17; extractives, 2; fat, 1; mineral matter, 2½.

The whole of the food fed to birds in the course of cramming is of a semi-liquid nature, and it is probable, therefore, that the process chiefly represents bringing up the flesh tissues to their maximum point of saturation; but, as this makes the birds better fitted for the table, this treatment is more than justifiable. The flesh of the Farne Island birds, from an edible point of view, was excellent.

During the Farne Island experiment 518 lbs. of food were used, which cost 29/9, or about 1⅜d. per lb.

During the Reading experiment 216 lbs. of food were used, which cost 14/9, or about 1⅝d. per lb.

The Reading food was therefore 18 per cent. dearer than the Farne Island.

Ninety per cent. of the Reading food was hard and 10 per cent. soft food.

Fifty-five per cent. of the Farne Island food was hard, and 45 per cent. soft food.

The Farne Island soft food was of a less, and the hard evening food of a greater, carbonaceous nature than the Reading. This enabled foods to be used which, whilst costing 18 per cent. less, were equal as flesh formers, and superior as bone and frame producers, and they were certainly as good, if not superior, from a health-maintaining point of view.

Therefore, owing to this superior nutritive balance of foods, the birds were able to eat a much larger quantity without any injury to their health just at the critical period of their lives—i.e., between the ages of from 4 to 13 weeks, when their health is the most delicate, and they are putting on flesh at the most economic rate.

By this means producers are enabled to hasten the growth and secure a quicker and more substantial profit from their birds.

The Farne Island birds had excellent health throughout. There were no signs of leg weakness or foundering, and yet they eat 97 per cent. more food, grew 93 per cent. faster, and their flesh cost 31 per cent. less to produce.

In order to show more clearly the effect of soft feeding, I have drawn a comparison between the lives of trees and chickens.

Trees grown on soft soil have an open and soft grain, and their timber is suitable for domestic purposes. Trees grown on hard soil have a close and hard grain, and their timber is suitable for heavy work. I have represented a chicken's life by concentric circles; each circle being its proportion of size at the end of each week, which corresponds to a tree's life in years (see next page).

The diagram was drawn on the scale of 1 lb. of

Diagram showing Chicken Growth.

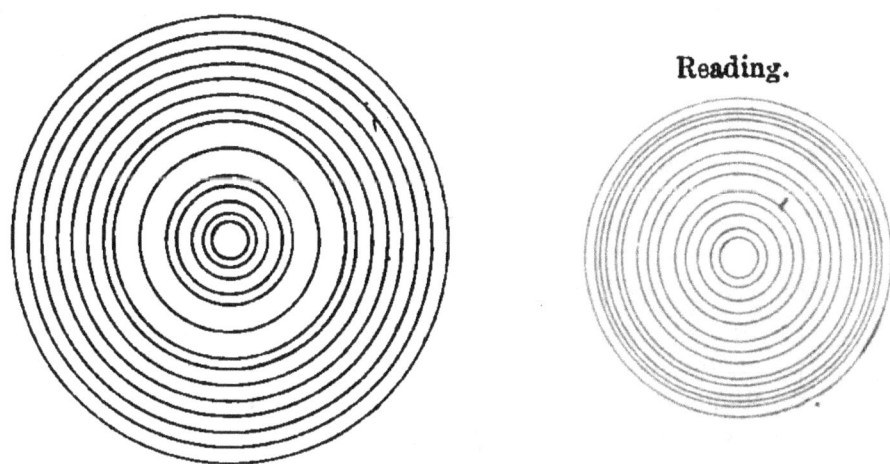

Farne Island.   Reading.

Drawn to scale, one pound to one square inch, and reduced together before engraving.

Diagram showing Chicken Growth at Reading in 1905, and Farne Island, 1906.

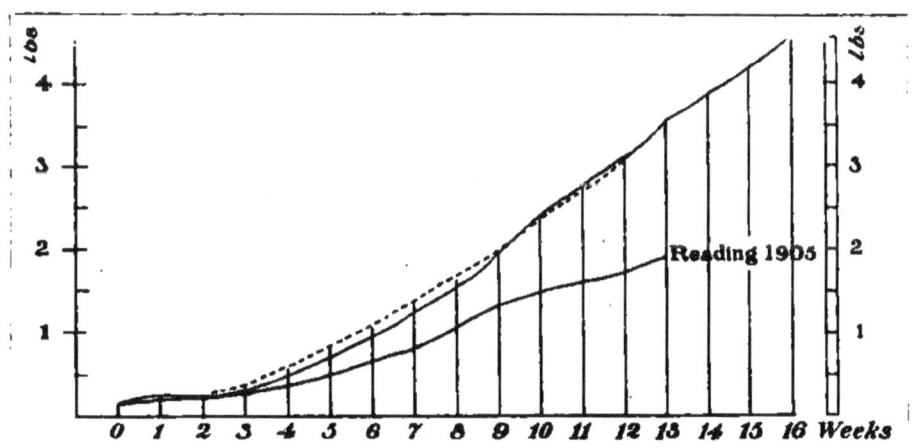

The dotted line shows the growth of the average chicken of the second brood in the Farne Island experiment, 1906, taken to the 12th week. The lowest ascending line shows the growth of the average chicken in the Reading experiment, 1905, taken to the 13th week. The highest ascending line shows the growth of the average chicken of the first brood in the Farne Island experiment, 1906, taken to the 16th week.

chicken flesh to 1 square inch of surface. The drawing was reduced on engraving, but the proportions remain the same.

Briefly, the claim of superiority for the Farne Island system is as follows :—

1. To feeding by Table.
2. To the last grain feed being given at sunset instead of 5 p.m.
3. To the first soft feed being given at 9 a.m. instead of 7 a.m.
4. To the non-carbonaceous nature of morning food and the carbonaceous nature of evening food.
5. To the superior nutritive balance of foods used.
6. To the use of bran (kept dry) and the exclusion of meat.
7. To the much larger proportion of soft food used, 45 per cent. against 10 per cent.
8. To the use of cheap and wholesome foods only.
9. To the expansion of the birds' crops, by feeding two full meals per day.
10. To the practice of giving never less than four meals per day (hard and soft alternately).

CHAPTER IV.

AN INTERESTING DEDUCTION ARRIVED AT FROM AN ANALYSIS AND COMPARISON OF THE READING AND FARNE ISLAND FIGURES.

EVERY pound of the 201 lbs. of food eaten at Reading, which cost on an average $1\frac{3}{6}$d. per lb., produced $4\frac{7}{16}$ oz. of flesh. Every pound of the 578 lbs. of food eaten at Farne Island, which cost on an average $1\frac{1}{6}$d. per lb., produced $4\frac{8}{16}$ oz. of flesh. Therefore, although the Reading food was of a different nature, fed in 97 per cent. less quantity, and cost 18 per cent. more than the Farne Island food, yet, as far as flesh production was concerned, it gave practically the same results at equal ages.

This agrees with the view formed by me during my original investigations in 1902, which was that most of the meals in general use for feeding poultry yielded, from the point of view of egg or flesh production, practically equal results. Therefore, the expensive advertised meals, though costing perhaps three or four times as much, will not produce more than meals purchased from local dealers, which has been milled from grain grown in the immediate neighbourhood. This description applies to the middlings and bran used by me on the Farne Island.

I believe it to be an absolute necessity, in order to make egg or chicken flesh production profitable, that a large quantity of low-priced, wholesome food, of a well-balanced nutritive ratio, bearing an analysis as near wheat as possible, should be supplied to the birds.

The Farne Island figures relating to Brood No. 1 show that when a bird has arrived at a marketable size, namely, 4 lbs., it will have eaten exactly 4 lbs. of food for every pound of flesh made.

This conclusion was confirmed by Brood No. 2, which, when weighing $3\frac{1}{2}$ lbs., had eaten $3\frac{9}{16}$ lbs. of food for every lb. of flesh made.

At Reading chickens 13 weeks old, weighing $1\frac{3}{16}$ lbs., required $3\frac{9}{16}$ lbs. of food in 1905, and $3\frac{10}{16}$ lbs. in 1904.

I think we are therefore justified in taking the Farne Island and Reading results in this respect as being somewhere near correct.

The deduction can scarcely be disputed that, when a chicken reaches four pounds (a marketable weight), the quantity of food required to make the four pounds of flesh weight amounts to sixteen pounds. And in order to still further show that it is the quantity of food eaten rather than the quality which regulates this, I have drawn up the following comparison, which

HOW TO MAKE POULTRY PAY. 43

shows the total weight of food eaten per bird up to the end of the eighth, ninth, tenth, eleventh, twelfth, and thirteenth weeks at Reading and Farne Island, and have worked out up to the end of each week the weight of flesh actually produced by every pound of food eaten up to the end of that period.

READING.

Number of weeks old	8	9	10	11	12	13
Total number of pounds of food eaten per bird to date	$3\frac{1}{10}$	$3\frac{7}{8}$	$4\frac{7}{10}$	$5\frac{3}{10}$	$5\frac{8}{10}$	$6\frac{1}{10}$
Number of ounces of flesh produced from each pound of food eaten	$5\frac{3}{10}$	$4\frac{3}{8}$	$4\frac{4}{10}$	$4\frac{7}{10}$	$4\frac{9}{10}$	$4\frac{7}{10}$

FARNE ISLAND.

Number of weeks old	8	9	10	11	12	13
Total number of pounds of food eaten per bird to date	$4\frac{6}{10}$	$5\frac{3}{8}$	$7\frac{4}{10}$	$8\frac{1}{4}$	$10\frac{3}{8}$	$12\frac{9}{10}$
Average number of ounces of flesh produced from each pound of food eaten	$5\frac{8}{10}$	$5\frac{5}{10}$	$5\frac{3}{10}$	$4\frac{1}{4}$	$4\frac{9}{10}$	$4\frac{8}{10}$

And now that the above has been proved, the following list has been drawn up as a protection to those who may be following my system, in order that they may ascertain the approximate cost of

44 HOW TO MAKE POULTRY PAY.

chicken flesh if produced from the various advertised meals.

lbs. of meal			per cwt.	per lb.		per lb. for food.
4	advertised by a leading firm	at 20/-	= -/2⅛		will be required to produce 1 lb. of flesh, therefore this flesh will cost . .	-/8½
4	,, Norwich ,,	at 19/-	= -/2			-/8
4	,, Birmingham ,,	at 18/-	= -/1⅞			-/7½
4	,, American ,,	at 14/-	-/1½			-/6
4	,, Orpington ,,	at 15/-	= -/1⁴⁄₇			-/6½
4	,, Poole ,,	at 11/9	= -/1¼			-/5
4	,, Reading ,,	at 7/9	= -/⅞			-/3¼
4	,, Farne Island ,,	at 5/8	= -/0⅝			-/2½

In order that this may be more clearly seen, I have reproduced (p. 48) the diagram shown in Chap. III. From it anyone can ascertain the balance of flesh which will be left to represent clear profit, after the food eaten and all charges have been paid for.

As an instance, foods costing on an average 2⅛d. per lb. are represented by the top thick line, and the top dotted line shows the additional flesh necessary to pay for the labour, rent, and all charges on a 4½ lbs. chicken.

I have gone to considerable trouble in order to ascertain the mean average values of the chemical constituents of the foods as used by Reading and by myself. They are as follows:—

	Protein	Fat	Carbon	Bone Making	Water	Fibre	Total
Reading . .	12¼	5	65	2¼	13½	2	100
Farne Island.	14½	4½	58½	4¾	12¾	5	100

HOW TO MAKE POULTRY PAY. 45

Theoretically the percentages of the various groups required to form a perfectly balanced nutritive ratio of foods should be arranged as nearly as possible to the first column given below.

	Perfectly Balanced Theoretical Ratio.	The Reading Ratio.	The Farne Island Ratio.
Nitrogenous ..	13½	12¼	14½
Carbonaceous ..	60	72	68
Salts	2½	2¼	4¾
Water	24	13½	12¾
	100	100	100

It is interesting here to compare what one might term the anticipated laboratory results with the actual results: in other words, to compare the weight of flesh-forming material actually eaten by the birds with the actual weight of the flesh formed.

	Total Weight of Food Eaten.	Total Weight of Flesh Produced.	Percentage.
	lbs.	lbs.	
At Reading ..	201	53	26
At Farne Island .	518	141	27

Therefore we arrive at the following comparison between practice and theory :—

	READING.		FARNE ISLAND.	
	THEORETICAL.	ACTUAL.	THEORETICAL.	ACTUAL.
Flesh producing constituents of foods eaten and the actual weight of the flesh produced	per cent. 21½	per cent. 26	per cent. 27	per cent. 27
Carbon	65	60	60¼	60¼
Water	13½	13½	12¾	12¾

If I am right in my deductions, then hundreds of thousands of pounds are annually being paid by people all over this country for advertised poultry meals, costing on an average from 12s. to 20s. per cwt., whereas meals costing perhaps only 5s. 8d. per cwt., which are equal to these, if not superior, in every respect, might be bought by poultry raisers from their local corn dealers.

The food bill, combined with proper management, will make all the difference between profit and loss.

During the course of my practical experiments, I took no notice of theory whatsoever. I simply passed the various foods through the birds in different ways, and eagerly watched the results. I had, as it were, two gauges at work; one—the commercial test—was applied to the actual cost of the flesh produced, week by week; and the other—the hygienic test — was applied to the birds' droppings, and their appearance, as an indication as to how the various foods and methods of feeding acted on their general health.

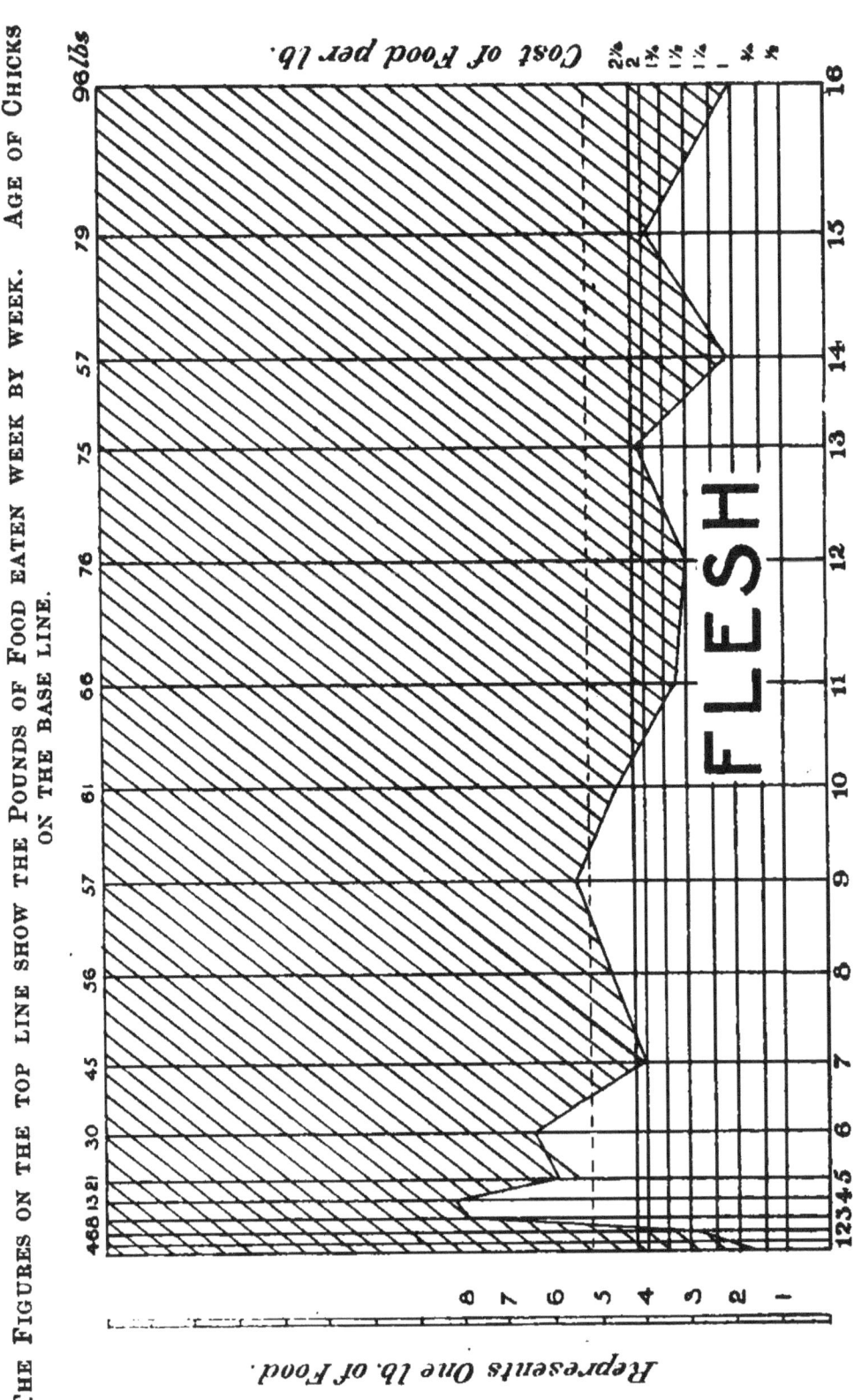

CHAPTER V.

APPLIANCES, FOODS, ETC., USED IN THE FARNE ISLAND EXPERIMENT.

IN March, 1906, I purchased the appliances necessary for the experiment, and, in order to prove my incubator and foster mother, made a small initial test on the mainland.

On May 16th I went across to Farne Island, off the Northumberland coast, so that I might make the complete experiment alone and undisturbed.

Although I landed on May 16th I was not able to get my appliances and eggs across, owing to bad weather, etc., until June 11th. My experiment then commenced, and continued without interruption until October 22nd.

I did not, however, leave the island until the beginning of December, as I wished to spend a month trying experiments of a minor nature which I was unable to undertake during the course of my large experiment without interfering with its progress.

I do not attribute my success to the period of the year in which the experiment was made, nor to the locality, "unstaled" ground, or climatic influences. I attribute them entirely to the food used and to the general system of feeding, watering, housing and management. I am confident that between the months of February and October equal, if not better, results can be reproduced practically anywhere.

During the months of November, December, and January, the days being shorter, the birds naturally will eat less food and take longer to grow, but the cost of the flesh for food will be about the same, though the labour and other charges may be slightly increased.

Labour.

During the last four months I was living quite by myself, and concentrated my whole thoughts and attention entirely undisturbed upon my work.

Location.

Inner Island, Farne Island. Area, 7 acres of grass. The birds had free range over the whole island. There are no trees, shrubs, hedges, or any natural shelter whatsoever. The drinking water was collected from the roof of the dwelling house.

Management.

Eggs were placed in the incubator immediately the preceding lot was hatched off. This was continued without interruption, and at the end of the fifteenth week I had under experimentation five broods containing 185 chickens of all ages, ranging from 1 to 15 weeks, as follows: 41 chickens, 15 weeks; 31 chickens, 11 weeks; 40 chickens, 8 weeks; 32 chickens, 5 weeks; and 43 chickens, 1 week old.

A fresh programme of experimentation was mapped out on each Saturday for the following week; sometimes if the results of a test did not appear sufficiently clearly defined it was continued over two weeks. Each individual experiment started at 12 on Monday morning and finished at 12 on the following Monday.

The birds were weighed every Monday morning at 12. Until the sexes could be distinguished 24 birds were weighed haphazard from each lot. As soon, however, as the sexes could be distinguished 12 cocks and 12 hens were weighed; a record was kept of the largest bird in each lot week by week.

No foods were used but those which can be purchased generally, and no attempt was made to force growth; the birds were treated by ordinary methods throughout.

The price of the flesh produced week by week by the various mixtures of foods and methods of feeding was carefully worked out and noted. The health of the birds and the effects of the different foods and methods of feeding on them were carefully watched, the chief indications being their general behaviour, their droppings, and the appearance of their plumage. All birds dying during the experiment were immediately opened, and the cause of death ascertained when possible.

Experiments on the effect of improper ventilation in foster-mothers were tried and noted.

The Incubator.

The incubator used was the Cypher, 125, Finsbury Pavement, London, 1906 pattern, 140 eggs size, hot air, cost £5.

The Foster-mothers.

Indoor Foster-mother.—The Incubator Component Parts Company, Stonehouse, Gloucestershire, 60 chick size, hot air, cost £1 1s. The following alterations were made: The wire netting was removed from the front portion and glass substituted. A large box was placed on legs and raised alongside, so as to form an open run for scratching purposes. A doorway was cut in the front to allow the chickens to enter this box. Two ventilation holes were bored, one on either side of the glazed portion. The total cost of alterations would be about 3s.

Outdoor Foster-mother.—Cypher model, Mountain Bailey, Woking Village, Surrey, 60 chick capacity, cost £3 15s., hot air.

HOW TO MAKE POULTRY PAY. 53

Cold Brooder.

I designed and made my own cold brooder. It is in four loose parts: sides, bottom, and 2 lids; it has excellent ventilation, and there is no draught.

Houses.

The houses are exactly the same as the cold brooder; they are fitted with a removable loose shutter at each end which lifts out. They stand touching one another, end to end; each separate apartment is intended to hold 16 chickens, 16 weeks old.

When the small chickens are first brought from the foster-mother only one apartment is used; when they are 9 weeks old the loose shutter is removed and the chickens then have the run of two houses, and when 12 weeks old, if there are sufficient chickens, a third is added. There is a sliding glass shutter in front which gives both light and ventilation. This type of house is extremely cheap, is perfectly lighted, has excellent ventilation, there are no draughts, it is easily cleaned, and can be taken to pieces in a moment. The final satisfactory results of the experiments were in a great measure due to their use. A small permanent stone shed on the island, 6 feet by 4 feet, was also utilised.

Feeding and Watering Appliances.

Indoor Foster-mother.—Watering: 1 of Lathrop's chick device, 1s., R. A. Colt, 73A, Chiswell Street,

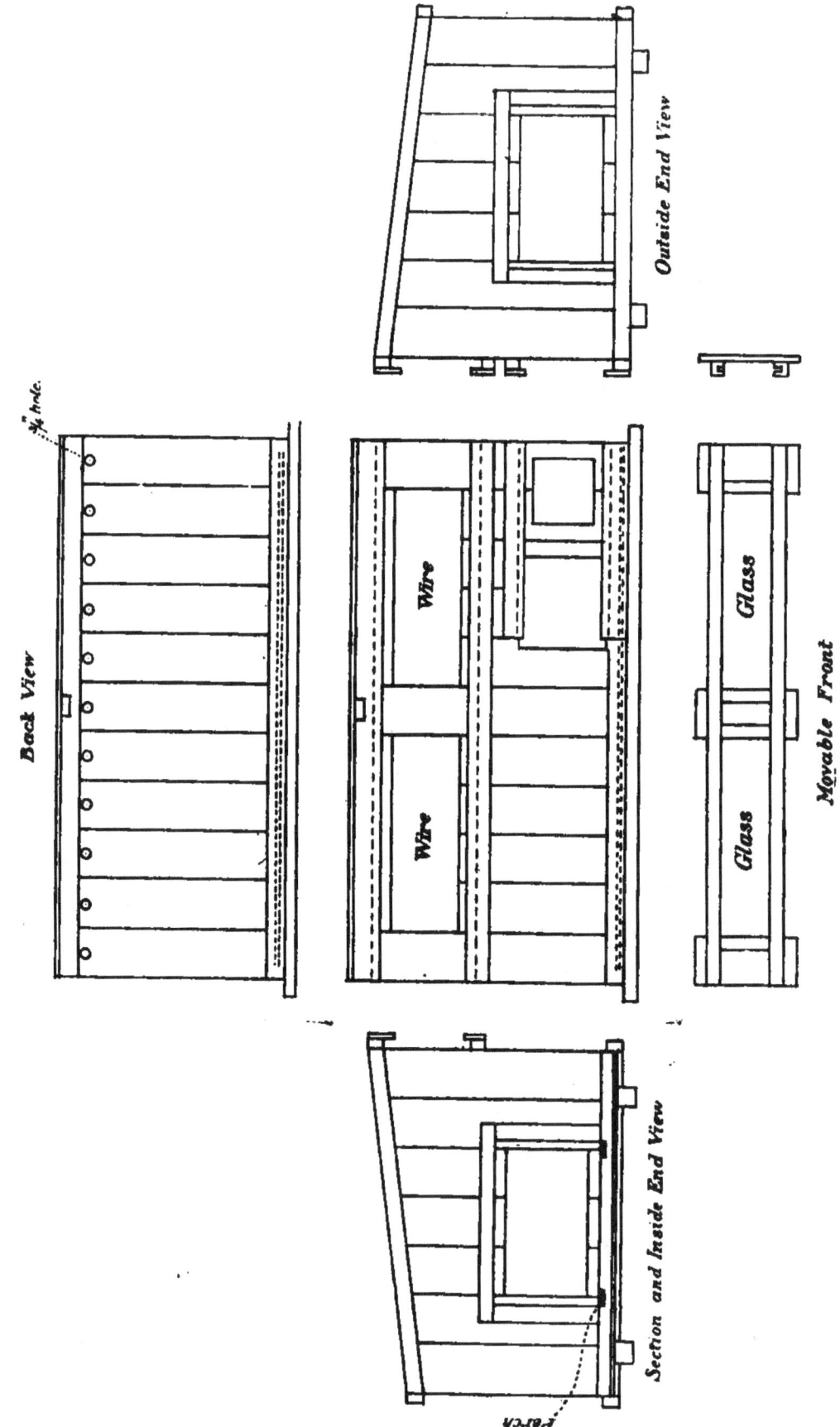

HOW TO MAKE POULTRY PAY. 55

London, E.C. Soft food: 2 of Tamlin's guarded feeding troughs, 1s. each, Messrs. Tamlin, 37, Sheen Road, Richmond.

Outdoor Foster-mother.—Watering: Inside, 1 of Lathrop's, as above. Outside, 1 4-quart rational fountain, cost 2s. 3d., R. Meech, Poole, Dorset. Soft food: Inside, 2 of Tamlin's, as above. Outside, 2 baking pans 8 in. by 4 in., cost 6½d. each, bought from Messrs. Gamage, Holborn, London. Grit: 1 baking pan, as above.

Cold Brooder.—Watering: 1 12-in. galvanized wash up, cost 6½d. Soft food: 2 12-in. galvanized wash up, cost 6½d. Grit: 1 12-in. galvanized wash up, cost 6½d., Messrs. Gamage.

Houses.—Watering: 1 galvanized 12-in. wash up, cost 6½d. Soft food: 6 galvanized 12-in. wash up, cost 6½d. Grit: 1 galvanized 12-in. wash up, cost 6½d., Messrs. Gamage.

Cost and Preparation of Foods.

The experiments lasted twenty-five weeks. I was thus able to utilise information as it was gained, and obtained more and more satisfactory results as the experiment progressed. The foods I finally selected, and my methods of preparing them, are set out on next page. I found that a thorough mixing of the soft foods had a decided effect on the birds.

A.—Dry Chick Feed.

Wheat	3 parts by weight.
Canary seed	2 ,, ,,
Oatmeal	2 ,, ,,
Millet	1 ,, ,,
Broken maize	1 ,, ,,
Buckwheat	1 ,, ,,
Rice	1 ,, ,,
Meat	1 ,, ,,
Grit	1 ,, ,,

Cost, 9s. 4d. per cwt., or 1d. per lb.

B.—Chick Soft Food.

Biscuit meal	6 parts by bulk, at	12s. 10d. per cwt.
D chicken meal	2 ,, ,,	5s. 8d. ,,
Rice	2 ,, ,,	10s. ,,
Oatmeal	2 ,, ,,	13s. 4d. ,,
Meat	1 ,, ,,	13s. 5d. ,,

Cost, 11s. 5d. per cwt., or $1\frac{3}{10}$d. per lb.

C.—Chicken Corn.

Wheat	25 per cent. by weight.
Barley	25 ,, ,,
Maize	20 ,, ,,
Dari	10 ,, ,,
Buckwheat	10 ,, ,,
Tares	10 ,, ,,

Cost, £6 10s. per ton, or 6s. 6d. per cwt., or $\frac{11}{16}$d. per lb.

D.—Chicken Meal.

Middlings	3 parts by weight, at £5 15s. per ton.
Coarse bran	1 ,, ,, ,, £5 10s. ,,

Cost, £5 13s. 10d. per ton, or 5s. 8¼d. per cwt., or $\frac{5}{8}$d. per lb.

How to Use Bran.

Great care must be exercised with the use of bran. It must be stored in a dry place, as it has a great affinity for moisture. If allowed to become damp it will slightly ferment, and when fed will cause bowel troubles; but if carefully treated and thoroughly mixed with the correct proportion of good sound middlings, it makes an excellent feeding material, and forms a splendid frame- and bone-maker.

Meat meal and bone meal appeared to me to act on the birds' systems much in the same way as the bran, but bran is much cheaper and just as good.

The middlings must be of a good sample and unadulterated. It will be found that, if purchased from an old-fashioned country mill, it will probably be of better feeding quality than if procured from one fitted with more modern machinery. Some middlings have practically no feeding value whatsoever.

The *B chick soft food* is to be prepared as follows: Just sufficient quantity for the one day is to be prepared every morning. The rice and meat are boiled together for ten minutes. The oatmeal is soaked in an equal bulk of boiling water for not less than a quarter of an hour. The water strained off the meat and rice is to be used for this purpose. The biscuit meal and chicken meal are mixed together dry, and then the whole of the ingredients are thoroughly mixed.

The *D meal* must be mixed with boiling water, great care being taken that the water is *actually boiling*. Four parts of meal are used to 1⅛ part of water. It must not be mixed in bulk, but in small quantities. An ordinary kitchen 3-prong fork was found to act admirably for this purpose.

A curious fact, which I could not disprove, was the superiority of bran mixed with middlings alone without any bone meal or meat meal added.

On two or three occasions I tried feeding chickens with bran and middlings, in competition with others, giving, in addition, a proportion of meat, and in every case the cost of the flesh produced by the bran-fed chickens would remain somewhere near constant, *i.e.*, anything between 2½d. and 3d. per lb. live weight, whereas the cost of flesh produced by those given the meat would run up to 4d., 5d., and even 6d. per lb. live weight.

In outward appearance the meat-fed chickens looked, if anything, healthier than the bran-fed, their feathers were more glossy, and their droppings somewhat more open.

This is quite contrary to what I was taught, and to what one learns from text-books, and to what is generally accepted as a fact.

I was particularly interested in this, as I have never yet met with a specialising poultry farmer who did not use meat in some shape or form; and, on the other hand, I do not know of a single ordinary farmer who either believes in it or uses it.

In the above I may be right or I may be wrong, but until I have it conclusively proved to me, that after the earliest stages, that is, up to the time the chicks leave the foster-mother, animal food artificially fed to them is essential, I shall not use it.

The Question of Grit.—Although I used grit throughout my experiments, because I had been taught that it was essential, nevertheless I am ignorant of the reason why it is essential.

In order to ascertain, if I could, what action the grit had, I opened the crops of every chicken that died, but never found the slightest indication of any grinding having taken place. The grains seemed to be slowly dissolved by the action of fluids, and from my observations I can scarcely credit that the grit, as is generally supposed, acts instead of teeth to grind the food. Notwithstanding this, however, I believe it is of use, though in some other way which I do not understand. Young turkeys require grit, if anything, more than young chickens.

Unfortunately I had no rock salt on the island, otherwise I should have broken some up small and mixed it with the grit, so that the chickens might help themselves. I believe this would have been beneficial to them, and assisted their growth.

I do not know if any extra precautions were taken at Reading with regard to the water, but on this point

I was most careful. I always had large quantities in deep open pans kept before the birds, close to their feeding troughs.

At nine weeks of age an average chicken should eat 3 ozs. of dry food per day, therefore they ought to drink 12 ozs. of water per day; and, allowing for 4 ozs. being fed to them with their soft food, this leaves 8 ozs. to be drunk at other times by each bird throughout the day. Therefore 40 chickens will require 5 pints of water, and they ought consequently to have placed before them not less than 10 pints of fresh water each morning.

Gapes is caused by impure water. If there is any sign of this disease, the water must be boiled, and a good supply should always be available.

CHAPTER VI.

PARTICULARS OF THE FEEDING AND GENERAL MANAGEMENT OF CHICKENS ON THE FARNE ISLAND SYSTEM.

From the day the chickens are removed from the incubator to the time they are marketed they are fed and managed on definite lines.

From one to fourteen days old the chicks are fed on a limited amount of dry chick feed, given them by Table, in small quantities every two hours.

From the tenth to the fourteenth day they have at 9 a.m. and 3 p.m. an additional feed of as much of the D chick soft food as they will eat. After half an hour any that is left is removed.

During the whole of the time they are in the outdoor foster-mother they have five regular meals per day, at daybreak, 9, 12, 3, and sunset.

The daybreak, 12, and sunset meals consist of grain; the 9 and 3 o'clock meals of soft food.

At 9 o'clock they have as much soft food, and at sunset as much grain, as they will eat; but at daybreak and 12 just half the amount of grain as is eaten at sunset; and at 3 half the amount of soft food as is eaten at 9 is fed to them.

From the time they leave the foster-mother until marketed they have four regular meals per day, at 9, 12, 3, and sunset.

After leaving the foster-mother the birds have one long period every 24 hours without food, the object being to clear their systems. Their first meal is given at 9 a.m. They therefore spend their early mornings hunting about and picking up grubs, worms, etc. This exercise in the cool of the mornings is extremely beneficial to them, and prevents bad effects, such as foundering, arising from their heavy meals during the remainder of the day.

At 9 the birds' crops are fairly empty, they are then given as much soft food in a damp, hot, crumbly condition as they will eat. They therefore completely fill their crops with a food which slightly swells after eating, and this in time tends to stretch their crops and make them larger. At sunset their crops are again well filled with whole grain.

By above means the birds' crops are gradually expanded, and they are thus enabled to take in a larger amount of soft food in the morning and hard grain at night.

The early morning and 12 o'clock meal of grain

is crushed, so that it may more quickly pass into their system; the evening meal is uncrushed, so that it may be assimilated slowly.

The morning soft food is of a low carbonaceous nature; this permits the use at night of a larger proportion of the cheaper grains, such as barley and maize, which are highly carbonaceous, and which could not otherwise be fed to the chickens in such large proportions without risk of injury to their health.

The birds are never allowed to experience extreme hunger.

The drinking pans are deep and large, and therefore not liable to become affected by the sun's rays. The water is never allowed to become dirty or stale, and the pans are washed out every morning and refilled with fresh pure water.

The houses are kept scrupulously clean, are well lighted, littered, and ventilated; sunny corners are arranged where the birds can sun themselves and scratch. The chickens of various broods are kept apart, so that the older ones may not interfere with the younger. The birds are well looked after in every respect; they are always clean, well fed, and contented, and are never allowed to become dirty, mopy, and miserable.

There is always a dry place for their feet, and this, although a small item, is a very important one, especially in a damp locality or on a cold clay ground.

How to Prepare and Feed the 9 o'clock Meal.

1. Refer to the Feeding Table. Measure out into separate lots the correct quantity of dry meal for each brood.

2. Mix each lot separately with boiling water.

3. When all is ready mixed take out quickly and distribute to the various broods.

4. Whilst doing this pick up and return to the mixing room the empty tins left from the previous soft food meal and place away ready to hand for the next meal.

5. Half an hour after this meal has been fed walk round, and against any brood that has eaten everything make a note that they are to be given an additional portion at their corresponding meal on the following day; if any food should be left make no alteration.

6. Return at once to the mixing room and make the necessary alterations to Feeding Table.

The same method is adopted at the sunset meal of grain, but in this case there is, of course, nothing to mix and no tins to pick up. The birds are given as much as they will eat, and any grain that is left after twenty minutes is at once returned to the mixing room.

At 12 and 3 the correct amount of food to give will be ascertained by reference to the Table; this will be just half the quantity that was eaten at 9 and sunset. It will be found that the birds will quickly dispose of this quantity.

HOW TO MAKE POULTRY PAY. 65

The Feeding and Management of Chickens from 1 to 14 Days Old whilst in Indoor Foster-mother.

The following table of the quantities of dry chick feed to be given to chicks from 1 to 14 days old was carefully drawn up from observations lasting many weeks.

The chicks are just slightly underfed; they therefore have to work from early morning until night in order to obtain sufficient food to satisfy their hunger. They can never over-eat themselves, and are always kept, more or less, in a state of excitement all day hunting about. The last thing at night the food is measured out for the following day, and a small quantity of it is scattered amongst the litter after dark ready for them at daybreak. The remainder is given in small quantities every two hours or so during the daytime.

Table giving the Correct Amount of Dry Chick Feed to be Fed to Chickens from 1 to 14 Days Old.

First day, one half-gill to every 14 chicks.
Second ,, ,, 12 ,,
Third ,, ,, 10 ,,
Fourth ,, ,, 8 ,,
Fifth ,, ,, 7 ,,
Sixth ,, ,, $6\frac{1}{2}$,,
Seventh ,, ,, $6\frac{1}{2}$,,
Eighth ,, ,, $6\frac{1}{2}$,,
Ninth ,, ,, 6 ,,
Tenth ,, ,, 6 ,,

On the eleventh day the chicks are given, at 9 and 3, as much of the B soft food mixture as they

F

will eat up entirely. The A feed is still given them in the same manner and quantities as formerly—half a gill to six chicks. This method continues until the fourteenth day, when the chicks are removed to an outdoor foster-mother.

Management whilst in Indoor Foster-mother.

The temperature of the sleeping apartment when the chicks are first placed in this foster-mother must be kept as near 90 degs. as possible, and be gradually reduced to about 80 degs.

First day.—The ventilators are kept closed, the lid right down, and a slight sprinkling of litter placed under the glass portion.

Second day.—The ventilators are opened and a little more litter added.

Third day.—The glass lid is lifted and rested on sticks.

Fourth day.—The chicks are now allowed to go into the exercise-room, and a much larger quantity of litter is placed here. This portion will now be used for scratching purposes only, whilst the glazed part will be used for watering, and after the tenth day the soft food should be fed there.

System of Feeding and Management of Chickens in the Outdoor Foster-mother.

Feeding.

At 9 a.m. as much of the B chick soft food mixture is given them, and at sunset as much of the A dry chick feed, as they will eat. At 12 just half the amount of A as is eaten at sunset is given, and at 3 p.m. just half the amount of B as is eaten at 9.

After dark the same amount of A as is fed at 12 is scattered amongst the litter in the exercise-room, ready for them at daybreak next morning. About a week before the chicks are to be removed from the foster-mother the more expensive ingredients in B are gradually reduced, so that when they leave this feed consists of the D meal only.

Grit of a suitable size is always kept before them in a shallow tin.

Management.

When the chicks are first placed in the outdoor foster-mother the temperature of the sleeping compartment must be kept as near 80 degs. as possible, and gradually reduced to about 70 degs.

Should it be sunny and warm during the daytime the lamp may be put out.

The exercise-room must be used for scratching purposes, and the warm chamber for resting, watering, and feeding the soft food in.

During the first two days the chicks must not be allowed outside the foster-mother, but on the third day, if warm and dry, they must be allowed outside into a small movable run, about 12 feet square. After two days this run enclosure should be removed, and the chicks given perfect freedom. However, until they are three weeks old they will not be allowed to choose their own weather—that is, if it is raining or very damp they must be kept in.

When considering the use of the outdoor foster-mother the year may be divided into three periods: summer (May, June, July, August, and September), spring and autumn (March, April, October, and November), winter (December, January, February).

During the first period the chicks may safely be removed from the artificial heat when from 3 to 4 weeks old.

During the second period the chicks may safely be removed from the artificial heat when from 4 to 6 weeks old.

During the third period the chicks may safely be removed from the artificial heat when from 6 to 8 weeks old.

When purchasing foster-mothers, if the object is to grow birds quickly for commercial purposes, it is important that this fact should be appreciated and remembered, as it will be readily understood that a foster-mother capable of efficiently holding 70 chickens during the first period might only be large enough for 50 during the second, and 30 during the third. Crowded chicks suffer from lack of oxygen, and the results of insufficiency of foster-mother capacity are a stunted growth and a high rate of mortality.

System of Feeding and Management of Chickens from the time they leave the Foster-mother until Marketed.

At 9 a.m. the birds are given as much of the D soft food as they can eat, and at sunset as much of the C chicken corn as they can eat; at 12 just half the amount of C as is eaten at sunset is given; and at 3 p.m. just half the amount of D as is eaten at 9 is given.

The 12 o'clock grain feed is crushed through an ordinary kibbling mill, but as soon as the birds are large enough to eat it the sunset meal is given uncrushed.

Suitable sized grit is always kept before them in shallow tins.

Management.

When the chickens are taken from the foster-mother they are placed in a cold brooder—a description of which is given in the previous chapter.

After three weeks a similar cold brooder is joined on, and when the chicks are 12 weeks of age a third if necessary. At this age perches are given them, as it was found that they grew quicker if allowed to sleep side by side on perches than when huddled together on the floor.

These perches are arranged in the houses longitudinally, so that the birds on entering walk right along from end to end without disturbing those that have already gone to roost.

Programme of a Day's Work in carrying out the Farne Island System.

Start sufficiently early to have the first meal ready by 9. Place the water on to boil to be used in mixing the 9 o'clock meal of soft food. Prepare sufficient of the B chick soft food to last all day. Refer to the Feeding Table and measure out the D soft chicken food in separate lots. Mix up this soft food with the water, which will now be boiling.

9 a.m.: Distribute this meal, which will now be ready prepared, in separate lots as quickly as possible. Pick up and return to the mixing room the empty tins from the preceding afternoon's meal. Turn the eggs in incubators. Trim, refill and replace incubator and foster-mother lamps. Clean and refill all the water pans with clean, fresh water. Walk round and take notes of the amount of food eaten; this should be done within half an hour of the time the food was distributed. Return to the mixing room and alter the Feeding Table ready for next

morning. Clean out the foster-mothers, brooders, and houses. The day's work is now practically over.

12.0: Refer to the Feeding Table, measure out separately and distribute the mid-day meal of grain.

2.30, or earlier if necessary: Refer to the Feeding Table, measure out separately, and prepare the 3 o'clock meal of soft food.

3.0: Distribute this meal as quickly as possible.

Sunset: Refer to the Feeding Table, measure out separately the evening grain meal, and distribute it as quickly as possible. Twenty minutes later walk round and take notes of the amount eaten. Return to the mixing room and alter the Feeding Table ready for the next evening.

After dark walk round and distribute to the younger chickens their daybreak meal of grain, and, if necessary, close up the foster-mothers and brooders for the night. Then turn the eggs in incubators. During the spare time in the afternoon the corn must be crushed, and if straw chaff is used for scratching purposes this must be cut and stored away ready for use.

If the above system is followed out, one man, at £1, and one boy, at 5s. per week, should comfortably be able to handle 1,600 chickens, and send 100 chickens to market every week; and if the birds average, as they should, $4\frac{1}{2}$ lbs. each, this will represent

450 lbs. per week, which at a profit of 3d. per lb. yields £5 12s. 6d.

My system of working the incubator was as follows: First day, before placing the eggs in the incubator they were well washed in warm water; by so doing any cracked ones were discovered, and they were much pleasanter to handle later when turning. I started the incubator as early in the day as possible. On the second day the eggs were not touched. On the third day, about 10 in the morning, the eggs were turned for the first time. In the evening, after sunset, the operation was repeated. The method I adopted was to carefully remove the two centre rows, then gently roll the others down with the palm of my hand and return those removed, which now formed the two outside rows furthest away from the centre. This was continued morning and evening until the seventh day, when the first test was made, and then continued until the fourteenth day, when the second test was made, and then on again until the eighteenth. In the evening of this day the eggs were turned for the last time, and the front portion of hay removed. On the twentieth day the hatching started, and finished on the twenty-first. On the morning of the twenty-second day I removed the chickens to an indoor foster-mother, and then gave the incubator a thorough overhaul and clean out. I took no notice of ventilation, and by following above system over five consecutive hatches I averaged 79 per cent., hatching from

HOW TO MAKE POULTRY PAY.

fertile eggs as follows: 80 per cent., 78 per cent., 75 per cent., 81 per cent., and 82 per cent., which I considered perfectly satisfactory.

The temperature was started at 103 degrees and gradually increased to $104\frac{1}{2}$ at hatching time.

Brood No. 1.

Date.	Days Old.	Feeding Time.				Date.	Days Old.	Feeding Time.			
		9	12	3	S			9	12	3	S
July 2	1	—	—	—	3½	Aug. 27	57	14	6	7	13
,, 3	2	—	—	—	4	,, 28	58	15	6	7	13
,, 4	3	—	—	—	5	,, 29	59	15	6	7	13
,, 5	4	—	—	—	6	,, 30	60	15	6	8	13
,, 6	5	—	—	—	7	,, 31	61	15	6	8	13
,, 7	6	—	—	—	7½	Sept. 1	62	15	6	8	13
,, 8	7	—	—	—	7¾	,, 2	63	16	7	8	14
,, 9	8	—	—	—	7¾	,, 3	64	16	7	8	14
,, 10	9	—	—	—	8	,, 4	65	16	7	8	14
,, 11	10	—	—	—	8	,, 5	66	16	7	8	14
,, 12	11	1	—	1	8	,, 6	67	16	7	8	14
,, 13	12	1	—	1	8	,, 7	68	16	7	8	14
,, 14	13	1	—	1	8	,, 8	69	16	7	8	14
,, 15	14	1	—	1	8	,, 9	70	16	7	8	14
,, 16	15	3	1½	3	1½	,, 10	71	16	7	8	14
,, 17	16	4	2	2	4	,, 11	72	17	7	8	14
,, 18	17	4	2	2	4	,, 12	73	17	7	8	15
,, 19	18	4	2	2	4	,, 13	74	17	7	8	15
,, 20	19	4	3	2	6	,, 14	75	17	7	8	15
,, 21	20	4	3	2	6	,, 15	76	17	7	8	15
,, 22	21	5	3	2½	6	,, 16	77	17	7	8	15
,, 23	22	5	3	2½	6	,, 17	78	18	7	9	15
,, 24	23	5	3½	2½	7	,, 18	79	18	7	9	15
,, 25	24	5	3½	2½	7	,, 19	80	18	7	9	15
,, 26	25	6	3½	3	7	,, 20	81	18	7	9	15
,, 27	26	6	3½	3	7	,, 21	82	18	7	9	15
,, 28	27	6	4	3	8	,, 22	83	18	7	9	15
,, 29	28	6	4	3	8	,, 23	84	19	7	9	15
,, 30	29	7	4	3½	8	,, 24	85	19	7	9	15
,, 31	30	7	4	3½	8	,, 25	86	19	7	9	15
Aug. 1	31	8	4	4	8	,, 26	87	19	8	9	16
,, 2	32	8	4	4	8	,, 27	88	19	8	9	16
,, 3	33	8	4½	4	9	,, 28	89	19	8	9	16
,, 4	34	9	4½	4½	9	,, 29	90	19	8	9	16
,, 5	35	9	4½	4½	9	,, 30	91	19	8	9	16
,, 6	36	10	4½	5	9	Oct. 1	92	19	8	9	16
,, 7	37	10	5	5	10	,, 2	93	20	8	10	16
,, 8	38	10	5	5	10	,, 3	94	20	8	10	16
,, 9	39	10	5	5	10	,, 4	95	20	8	10	16
,, 10	40	10	5	5	10	,, 5	96	20	8	10	16
,, 11	41	11	5	5	10	,, 6	97	20	8	10	16
,, 12	42	11	5	6	10	,, 7	98	21	8	10	16
,, 13	43	11	5	6	11	,, 8	99	21	8	10	16
,, 14	44	11	5	6	11	,, 9	100	21	8	10	16
,, 15	45	12	5	6	11	,, 10	101	21	8	10	16
,, 16	46	12	5	6	11	,, 11	102	22	8	11	16
,, 17	47	12	5	6	11	,, 12	103	22	8	11	16
,, 18	48	12	5	6	11	,, 13	104	22	8	11	16
,, 19	49	13	6	6	12	,, 14	105	22	8	11	16
,, 20	50	13	6	6	12	,, 15	106	22	8	11	17
,, 21	51	13	6	7	12	,, 16	107	23	8	11	17
,, 22	52	13	6	7	12	,, 17	108	23	8	11	17
,, 23	53	13	6	7	12	,, 18	109	23	8	11	17
,, 24	54	14	6	7	13	,, 19	110	23	8	11	17
,, 25	55	14	6	7	13	,, 20	111	23	8	11	17
,, 26	56	14	6	7	13	,, 21	112	23	8	11	17

The figures under the hours of feeding stand for ¼-gills.
S stands for sunset.
From the first to the tenth day the chickens were fed every two hours, and the total daily amount given is entered in the above table under S.

CHAPTER VII.

THE AUTHOR'S SUGGESTIONS FOR AN EXPERIMENTAL POULTRY FARM. FACTS AND FIGURES FOR WOULD-BE POULTRY FARMERS.

I DESIGNED and suggested an experiment which should be carried out at an Agricultural Experimental Station, and explained it to a committee responsible for the management of such a station at their January meeting, 1907.

Outline of Experiment.

To put down a plant capable of producing 100 chickens per week, or, in other words, to create a machine which should have an output of from 450 to 500 lbs. of chicken flesh per week.

Primary object of the Experiment.

1. To determine the exact cost of producing chickens of marketable weight, *i.e.*, from 8 to 9 lbs. per couple.

2. To apportion to each chicken its share of the expenditure on eggs, oil, food, labour, and rent.

3. To determine the life of modern poultry rearing appliances (incubators and foster-mothers).

4. The consequent amount for depreciation and interest on above (to be charged against each bird).

5. To determine the economic feeding values of various home-produced and advertised foods.

6. To determine the flesh-producing qualities of various foods.

7. To determine the value of chicken manure on grass-land and for horticultural work.

Secondary objects of the Experiment.

1. To prove that poultry can be reared profitably in this country.

2. To show how to lay out capital on poultry appliances in order to yield the best commercial return.

3. To show how, by economic feeding and management, chicken flesh can be grown in this country so as to be sold and leave a fair margin of profit at 6d. per lb. live weight.

4. To show that if chicken flesh can be produced and sold profitably at 6d. per lb. live weight then the importation of foreign frozen birds would cease, as foreign birds cannot compete successfully with home-grown birds on our markets under this price.

5. To show that unless chickens are fed on home-grown and home-milled foods they cannot be produced profitably, the market price of chicken flesh being

regulated to a great extent by the market value of wheat in the particular district in which the chickens happen to be reared.

6. To show the impossibility of making chickens a commercial success if high-priced meals and grains are used.

7. To show that not only does the use of these foods produce flesh of an inferior quality, but further the birds are neither so healthy, active, or strong; they therefore grow slower, and are much less profitable than if fed on the cheaper home-grown and home-milled varieties.

8. To show that if chickens are fed on the American-grown and imported foods, and on many of the English high-priced advertised foods, the cost of production for food alone may be as high as $8\frac{1}{2}$d. per lb. live weight, instead of $2\frac{1}{2}$d. American and Russian frozen chickens are sold on our English markets as low as 6d. per lb. live weight; therefore, as long as English producers use these high-priced foods, the English public will buy frozen chickens. The Americans boom foods on our markets by the use of which chicken flesh cannot be produced and sold under 9d. per lb., and they are, in consequence, to some extent able to make a market in England for their own birds.

9. The birds produced by the experiment might be pure bred, and might be supplied to the farmers and others in the country at fair market prices.

Farmers could then obtain close at hand the best breeds for either flesh or egg production, at reasonable prices, to suit their soil, climate, and local markets. They could also obtain, close at hand, pure-bred cockerels for crossing purposes. The experiment, from this point of view alone, would be of great assistance to the poultry industry of any country.

On the basis of the facts ascertained in the Farne Island experiment, results of these investigations might be confidently expected, as follows:—

The maximum cost of production per chicken at a marketable age and weight, *i.e.*, about 16 weeks old and averaging 9 lbs. per couple, will be—

	s.	d.	
Per head for food	1	0¼	each, or 2$\frac{13}{16}$d. per lb. L.W.
,, for eggs and oil	0	3	,, $\frac{11}{16}$,,
,, for labour	0	3	,, $\frac{11}{16}$,,
,, for salesman's charges and carriage	0	3	,, $\frac{11}{16}$,,
,, for depreciation and interest on incubators and foster mothers	0	0¼	,, $\frac{1}{16}$,,
	1	9½	each, or -/4⅞ per lb.

The minimum average wholesale market value of above will be—

	s.	d.	
During autumn and winter	2	6	per head
During spring and summer	3	6	,,
Averaging	3	0	or 8d. per lb.

The foregoing shows an average profit of 1s. 2½d. per head all the year round, *i.e.*, a profit on 100 chickens per week of £6 0s. 10d., or £314 3s. 4d. per year.

Or after deducting the rent of say 10 acres	. . . .	£24
,, depreciation on houses, etc.	. . .	£13
,, ,, on laying hens	. . .	£ 3
		£40
Leaves a net profit of	£274 3 4 per year.	
Capital required (see page 82) . .	£253 5 0	

During the Farne Island experiment an incubator holding 72 eggs was started every three weeks, so that at the end of the nineteenth week there were five broods of chickens, ranging in number from 32 to 40, and in age from 3 to 16 weeks, under experimentation; or, in other words, there was an average of 36 chickens ready for successive marketing every 3 weeks, or 12 per week.

Now if the food bill averaged 12s. per week, I knew that each completed chicken was costing for food 1s., and I was thus enabled to obtain at a glance a fair estimate of the cost of producing a chicken.

The same idea was to be applied to my new experiment, though on a much larger scale. In this case an incubator containing 200 eggs was to have been started every Saturday. This would have required three incubators always working, and one to act as a go-between; and as this was to have been continued week after week without interruption, therefore on the nineteenth Saturday there would have been

200 eggs just placed in incubator.

400 eggs in course of incubation.

100 odd chickens just removed from incubators.

1,400 odd chickens of all ages, in broods of 100 odd, either in foster-mothers or running about.

100 chickens, 16 weeks old, to send to market.

And every following Saturday 200 eggs would have been placed under incubation, and 100 odd chickens, 16 weeks of age, sent to market. Now under the above plan, if the food bill had averaged £5 per week, then each completed chicken would have cost 1s. per head for food. If the labour bill had come to 25s. per week (one man at £1 and one boy at 5s.), then the labour charge would have been 3d. per head. If the eggs had cost an average of 1¼d. each, then this charge would have been 2½d. per head. If the oil bill had come to 4s. 2d., then this charge would have been ½d. per head. If the salesman's charges had come to 25s., then this charge would have been 3d. per head.

When once this, so to speak, living machine had been fairly started and satisfactorily working, then the economic values of various foods might be accurately tested, *i.e.*, the exact cost per lb. live weight of the flesh produced from feeding the various meals and grains might be ascertained, also the best methods, quantities, and times of day for feeding;

HOW TO MAKE POULTRY PAY.

also the periods of a bird's life when the flesh is produced in relation to the quantity of food eaten, at the most economic rate.

Now the above plan is just as applicable to poultry farming for ordinary commercial purposes as for a scientific test, and the foods to be used, the methods of feeding, and the results which will be secured from it are given in detail in the preceding chapters.

List of Appliances which will be required to turn out an average of 100 chickens per week:—

	£	s.	d.	£	s.	d.
Incubators—4 240-egg Machines	28	0	0	28	0	0
Brooders for chicks from 0 to 1 week old, two at 30/-	3	0	0			
Brooders for chicks from 1 to 2 weeks old, two at 30/-	3	0	0			
Brooders for chicks from 2 to 3 weeks old, two at £3 15s. 0d.	7	10	0			
Brooders for chicks from 3 to 4 weeks old, two at £3 15s. 0d.	7	10	0			
Brooders for chicks from 4 to 5 weeks old, two at £3 15s. 0d.	7	10	0			
Brooders for chicks from 5 to 6 weeks old, two at £3 15s. 0d.	7	10	0	36	0	0
Houses for chicks 6 to 7 weeks old, two at 45/-	4	10	0			
,, ,, 7 to 8 ,, ,, 45/-	4	10	0			
,, ,, 8 to 9 ,, ,, 45/-	4	10	0			
,, ,, 9 to 10 ,, ,, 45/-	4	10	0			
,, ,, 10 to 11 ,, ,, 45/-	4	10	0			
,, ,, 11 to 12 ,, ,, 45/-	4	10	0			
,, ,, 12 to 13 ,, ,, 45/-	4	10	0			
,, ,, 13 to 14 ,, ,, 45/-	4	10	0			
,, ,, 15 to 16 ,, ,, 45/-	4	10	0			
,, ,, 16 to 17 ,, ,, 45/-	4	10	0	45	0	0
1 Soft Food Mixing Machine	2	0	0			
1 Special Chaff Cutter	2	0	0			
1 Hot-water Apparatus	1	10	0			

	£	s.	d.	£	s.	d.
Brought forward	5	10	0	109	0	0
1 Sack Truck	1	10	0			
1 Weighing Machine	1	10	0			
1 Barrel of Oil, and Oil Filler	2	5	0			
Egg Boxes	1	0	0			
1 Testing Lamp	0	7	6			
1 Corn Grinding Machine	1	10	0			
Food Trays	5	0	0			
Water Pans	1	15	0			
Measuring Scoops and Mixing Pans	2	0	0			
Scratching Sheds	4	10	0			
Fowl Crates for marketing	2	10	0			
Peat Moss for Litter	1	0	0	30	7	6
Stock Birds—80 hens and 10 cocks, 90 at 3/6	15	15	0	15	15	0
Chickens—1,600, averaging 9d. each	60	0	0	60	0	0
Eggs—600 in incubators at 1¼d. each	3	2	6	3	2	6
Food—Two weeks' food in stock at £5	10	0	0	10	0	0
Labour—20 weeks at £1	20	0	0			
,, 20 weeks at 5/-	5	0	0	25	0	0
				253	5	0

	£	s.	d.
Plant	139	7	6
Labour	25	0	0
Chickens	60	0	0
Eggs, Incubating	3	2	6
Food	10	0	0
Stock Birds	15	15	0
	253	5	0

I strongly recommend hot-air incubators of a larger egg-carrying capacity than is actually required, as by this means better percentages of hatches from fertile eggs are obtained, and they are therefore, from the commercial point of view, well worth their small extra initial cost.

If the scheme is to be used for commercial purposes, I recommend, in place of keeping stock birds, that the eggs be purchased from people who make a business of collecting from cottagers and small farmers, in order to carry them to their local market towns. People of this class will be found in every rural district, and it was from these dealers that I obtained the eggs used by myself, and they gave perfectly satisfactory results in every way. Of course, if this plan is adopted, then the initial cost of the stock birds will be saved.

If the scheme is for scientific purposes, then I recommend that the stock birds be distributed in lots of about one dozen amongst cottagers or small farmers, and the eggs produced be purchased back from them at 1d. per dozen above the market value. Of course, the cottagers will find all labour and food, and the chickens will still remain the property of the experimental station.

I drew the attention of the Secretary of a County Council Technical Education Committee to the following: that after the machine was fairly established and satisfactorily running, it should be utilised for educational purposes. People who have lost their health in our large towns might be shown how they may return to the country, take up this branch of industry, regain their health, and make a living. I suggested to him that they should be given a course of instruction lasting sixteen weeks, and for this they should either pay a small fee, or the course should be free to deserving cases, in return for their services. By the end of sixteen weeks they would be fully qualified to take charge of a small machine for themselves. They should then have either rented to them or sold to them by the County Council a small machine, being a portion of, and being taken from the large mother machine, as follows:—

	£	s.	d.
2 140-egg Hot-air Incubators, at £5	10	0	0
2 Indoor Brooders, at 25/-	2	10	0
2 Outdoor Brooders, at 75/-	7	10	0
6 small Houses, at 45/-	13	10	0
Miscellaneous appliances	5	0	0
Food, Eggs, Oil, etc.	5	0	0
50 chicks 1 day old. 50 chicks 2 weeks. 50 chicks 4 weeks. 50 ,, 6 weeks. 50 ,, 8 ,, 50 ,, 10 ,, 50 ,, 12 ,, 50 ,, 14 ,, = 400 chickens, averaging - 9 each	15	0	0
	58	10	0

One incubator with 100 eggs would be started every other Saturday, and 50 chickens, 16 weeks old, would be ready for marketing once a fortnight, or 25 per week.

As there would be no labour charges, and there ought to be no marketing charges, provided the producer was energetic and worked up a local connection, he ought thus to be enabled to earn at once from 25s. to 35s. per week on an outlay of under £60.

I also pointed out to him that all workhouses, where practicable, should have a similar machine, which the inmates could operate themselves. They could then produce chicken flesh at the same price as they would be now paying for beef and mutton, and it would make a pleasant change of food, and be of benefit to their invalids.

CHAPTER VIII.

A COMPARISON BETWEEN CATTLE, SHEEP, AND POULTRY FROM A FARMER'S POINT OF VIEW—THE WEIGHT OF MARKETABLE CHICKENS — HOT AIR AND TANK INCUBATORS—CHEMICAL CONSTITUENTS OF FOODS—MAGNITUDE OF THE POULTRY AND EGG INDUSTRY IN THE UNITED STATES OF AMERICA.

A Comparison between Cattle, Sheep, and Poultry from a farmer's point of view.

CHICKENS reared at the Farne Island averaged when 16 weeks old, or 112 days, $4\frac{1}{2}$ lbs. or 72 ozs., which represents an average gain in weight of $\frac{5}{8}$ of an ounce per day.

The average increase in weight per day of cattle and sheep (taken from the reports of the Agricultural Experimental Stations) is represented by respectively $1\frac{1}{2}$ lb. and 4 oz.

Therefore from the point of view of flesh production:

> 1 beast will produce $1\frac{1}{2}$ lb. of flesh per day.
> 38 chickens ,, $1\frac{1}{2}$,, ,,
> 1 sheep will produce 4 oz. ,, ,,
> 6 chickens ,, 4 ,, ,,

HOW TO MAKE POULTRY PAY.

At the present time farmers are clearing a net profit on an average of about ½d. per lb. live weight on cattle, and ¾d. per lb. live weight on sheep.

I have shown that if chickens are reared and fed on the Farne Island principle they may be expected to yield an average profit of 3d. per lb. live weight. Therefore, if looked at from the point of view of financial investment, the relative values of cattle, sheep, and poultry are somewhere near the following:

1 beast, producing 1⅓ lb. of flesh per day, and representing a capital investment of approx. £10, yields a profit of ¾d. per day.

7 chickens, producing 4 oz. of flesh per day, and representing a capital investment of approx. 7s., yields a profit of ¾d. per day.

1 sheep, producing 4 oz. of flesh per day, and representing a capital investment of approx. 24s., yields a profit of $\tfrac{3}{16}$d. per day.

2 chickens, producing 1¼ oz. of flesh per day, and representing a capital investment of approx. 2s., yields a profit of $\tfrac{3}{16}$d. per day.

Therefore, £100, if invested in poultry, will yield a profit equal to that obtained from £2,800, if invested in cattle, at present prices.

Therefore, £100, if invested in poultry, will yield a profit equal to £1,200, if invested in sheep, at present prices.

Now the above comparison is valuable because one of the main difficulties in getting people back to the land is usually that of insufficiency of capital, and my experiments seem to prove that whereas ordinary farming requires a large capital and a comparatively small amount of attention. Poultry farming requires little capital but a great deal of personal labour and attention.

The Weight of Marketable Chickens.

I have seen the fact stated in periodicals, and in books, that a 13-week-old chicken should cost to produce for food on an average 5¾d. per head, as Reading had done this on a comparatively large scale. Now such a statement is misleading, because the Reading birds, which had cost 5¾d. at 13 weeks, only then weighed 29 oz. And a chicken of this weight, except under exceptional circumstances or fed for special market purposes, is not what one might call a commercial chicken, *i.e.*, it is neither saleable for table purposes or what one might term crammable, and therefore it is at a size which one might term neither fish, flesh, fowl, nor yet good red herring. Birds reared during the Farne Island experiment weighed 56 ozs. at 13 weeks, and had cost for food 8¾d. per head, and though still slightly on the small side they would have been saleable either for table purposes or for cramming purposes. As chickens at 13 weeks have passed their economic age for producing flesh therefore the Reading chickens will now cost (at the price of their food stuffs) to bring up to the Farne Island standard of weight, probably, instead of 5¾d. from 1s. to 1s. 3d. per head.

I recommend the following Appliances to those intending to carry out my System.

Sundry Appliances.—As per list given in Chapter V.

Drinking and Watering Appliances.—As described in Chapter V.

Houses.—As per drawing in Chapter V., and as described by me in that Chapter.

Cold Brooders.—Chapter V., as above.

Indoor Foster-mothers.—The incubator component parts, as described in Chapter V.

Outdoor Foster-mothers.—Those sold by R. A. Colt and the Allen Poultry Company are excellent, but for my particular system I strongly recommend the Cypher 1907 pattern *only*.

Incubators.—Cypher's latest pattern.

Much is claimed by the makers of tank incubators, yet from an industrial and investment point of view I recommend the hot-air machines supplied by The Cypher Incubator Company, R. A. Colt, The Allen Poultry Company, Incubator Components Parts, etc., and this view will be endorsed by thousands of people all over this country.

Although I have worked many tank machines Messrs. Cypher's 1906 pattern is the only hot-air machine with which I have had practical experience; but from this experience, and from the experience of friends who have worked other makes of hot-air machines, I have the greatest confidence in recommending this class to practical men as the *commercial* machine.

One of the best, if not the best make of tank incubator on the English market, which is recommended at many agricultural experimental stations, when constructed to hold 240 eggs costs £12 17s. 0d. Whereas the hot-air machine of equal capacity, similar in type to that used by me, costs £6 17s. 6d.

Reading obtained during the three years ending March, 1906, with tank incubation an average of hatchings of 74·44 per cent. of their fertile eggs. I obtained from my latest five consecutive hatchings with hot-air incubation average of 79 per cent. of my fertile eggs.

One of the greatest pitfalls in the poultry industry is that of over-capitalisation, therefore I have written the above with the idea of protecting those who may decide to follow my system with the object of making money out of poultry rearing.

Having no connection whatsoever with incubator manufacturers or any other vendors of poultry requisites, therefore my opinions are absolutely unbiased.

At Reading during the three years ending March 31st, 1906, an average percentage of hatchings were obtained from hot-air incubators of 71·36 of the fertile eggs, and during the same period from tank incubators 74·44 per cent., or 4·36 per cent. better.

Although I visited Theale in 1902 I did not then have an opportunity of seeing the incubator house, but from what I have heard and can gather I believe the hot-air machines there are being worked under somewhat disadvantageous conditions; they should be placed by themselves in a larger and more airy building. I feel confident that their hatchings from this class of machine might be improved.

I am under the impression that they are working with an insufficient supply of oxygen, and if the machines were operated by an experienced person, and the conditions were as favourably arranged as they could be for a test, which is to be a commercial one more than a scientific one, then the results obtained with the Cypher machines, from the point of view of the purchasing public, would be more favourable than those obtained by the tank machines.

An interesting experiment would be to take 50 chickens hatched in a hot-air machine, and 50 hatched in a tank machine, and to rear them under precisely similar conditions.

My own experience has shown that chickens hatched by hot-air machines were stronger, sturdier, and easier to rear than those hatched by the tank. The reason for this I have never been able to discover.

Chemical Constituents of Foods.

NAME.	PROTEIN.	FAT.	CARBON.	BONE MAKING.	TOTAL.	WATER.	FIBRE, SAND, ETC.	TOTAL.	GRAND TOTAL.
Meals.									
Oatmeal	18	6	63	2	89	9	2	11	100
Maize meal	11	8	65	1	85	10	5	15	100
Toppings	18	6	53	5	82	14	4	18	100
Middlings	18	6	53	5	82	14	4	18	100
Bran	16	4	43	18	81	13	6	19	100
Barley meal	11	2	60	2	75	11	14	25	100
Malt dust	24	2	44	7½	77½	10	12½	22½	100
Biscuit meal	12	2	75	—	89	10	1	11	100
Spratt's Patent Chicken Meal	17½	2	67	3	89½	9½	1	10½	100
Grains.									
Wheat	12	3	70	2	87	12	1	13	100
Dari	8	4	71	3	86	12	2	14	100
Maize	11	8	65	1	85	10	5	15	100
Millet	12	4	65	4	85	13	2	15	100
Buckwheat	15	3	64	2	84	14	2	16	100
Rice	7	—	80	—	87	13	—	13	100
Linseed	10	38	30	2	80	10	10	20	100
Beans and peas	25	2	48	2	77	10	13	23	100
Canary seed	10	5	60	2	77	13	10	23	100
Hemp seed	10	32	32	2	76	10	14	24	100
Barley	11	2	60	2	75	11	14	25	100
Tares	10	2	60	2	74	12	14	26	100
Oats	15	6	47	2	70	10	20	30	100
Miscellaneous.									
Potatoes	2	—	21	2	25	75	—	75	100
Milk	4½	3	5	¾	13¼	86¾	—	86¾	100
Meat scrap	16	—	18	4	38	62	—	62	100

Fresh-cut bone and horse flesh are rich in protein (about 20 per cent.). In dried fish the percentage of protein is still higher.

Average Cost of Food at Wholesale Market Prices.

Wheat, per quarter, 27/- £6 0 0 per ton, 6/0 per cwt., = -/1$\frac{8}{16}$ per lb.
Barley ,, 24/- £5 6 8 ,, 5/4 ,, = -/1$\frac{0}{16}$,,
Oats ,, 20/- £4 8 4 ,, 4/5 ,, = -/1$\frac{8}{16}$,,
Middlings £5 15 0 ,, 5/9 ,, = -/1$\frac{9}{16}$,,
Bran £5 10 0 ,, 5/6 ,, = -/1$\frac{9}{16}$,,

Average . £5 8 0 .. 5/5 ., -/1$\frac{9}{16}$

Magnitude of the Poultry and Egg Industry in the United States of America.

(*A statement by a well-informed American observer. Reprinted by permission.*)

The poultry and egg industry of the United States is fast assuming amazing proportions. Unfortunately, complete statistics of poultry and eggs are not obtainable. Statistical information concerning poultry and poultry products is only taken during census years. In the latest census, records were obtained showing the number of chickens and other poultry three months old and over on June 1st, 1900, and upon farms alone. Obviously, an enormously large number of young stock were not accounted for, neither was any record made of the immense output of squab broilers, broilers, and small roosters that are turned out in hundreds of thousands annually by large speciality poultry farmers. Green ducklings also are not included in these statistics, since the great majority of specially-grown market ducks are disposed of to furnish delectable morsels for epicurean palates while still at the tender age of nine or ten weeks.

Added to this are the hundreds of thousands of fowls and chickens, together with the eggs produced by them, that are grown on town and city lots, of which it has been impossible to obtain any record whatsoever. Were it possible to obtain even a fair estimate of these it is probable that the total valuation of poultry and eggs would be increased by many millions of dollars. Even on farms it has been

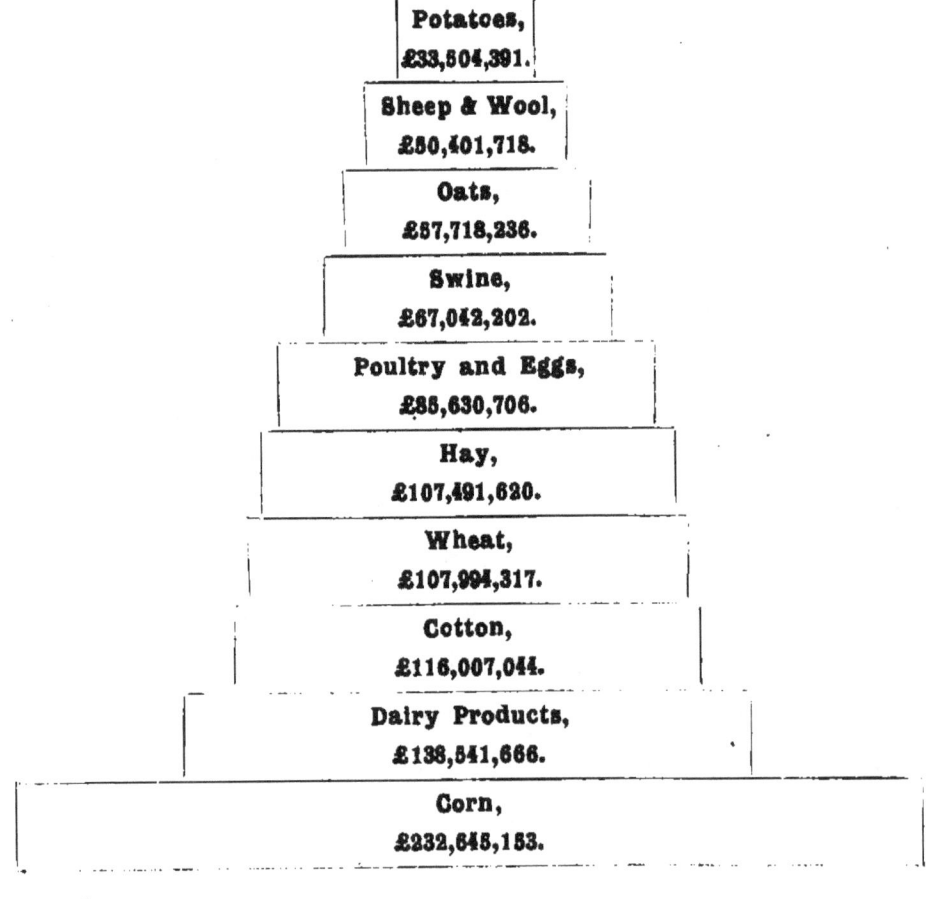

Potatoes, £33,504,391.

Sheep & Wool, £50,401,718.

Oats, £57,718,236.

Swine, £67,042,202.

Poultry and Eggs, £85,630,706.

Hay, £107,491,620.

Wheat, £107,994,317.

Cotton, £116,007,044.

Dairy Products, £138,541,666.

Corn, £232,645,153.

exceedingly difficult to obtain absolutely reliable statistics, because farmers are notoriously careless, as a general rule, concerning the records of their poultry and egg production.

The figures given in this chapter for the year 1905 are based on statistics supplied to us through the courtesy of the United States Department of Agriculture, and we have been obliged to follow the same plan that was adopted in 1900. Therefore in this statistical table no allowance whatever is made for the enormous production of squab broilers, broilers, small roosters, green ducklings, and town and city lot poultry products. Were reliable figures, representing this part of the poultry industry, available, the grand total valuation of poultry and eggs for the United States for 1905 would in all probability equal or exceed the value of the wheat crop and become a formidable rival of the cotton crop and dairy products.

www.ingramcontent.com/pod-product-compliance
Lightning Source LLC
Chambersburg PA
CBHW082343220526
45470CB00008B/2625